MANEATERS

MANEATERS

By
Peter
Hathaway
Capstick

PETERSEN PUBLISHING CO.

Petersen Publishing Co., 8490 Sunset Blvd., Los Angeles,
CA 90069

Library of Congress Catalog Card Number: 81-82705
ISBN 0-8227-3023-5

Printed in the United States of America

DEDICATION

To the memories of
Thomas Capstick
and
Thomas Capstick, Jr.

CONTENTS

CHAPTER ONE

Sharks 2

CHAPTER TWO

Lions 28

CHAPTER THREE

Crocodiles 64

CHAPTER FOUR

Leopards 80

CHAPTER FIVE

Bears, Wolves, Hyenas 96

CHAPTER SIX

Tigers 118

CHAPTER SEVEN

Man on the Menu 132

CHAPTER EIGHT

Questionable Killers 146

CHAPTER NINE

Cannibals 160

Illustrations by Leon E. Parson

INTRODUCTION

You would probably not have let him date your daughter; I doubt you would have lent him carfare. We don't know all that much about him except that he lived in the vicinity of a cave in South Africa that somebody named Swartkrans about a million years after something grabbed him one dark night, killed and almost surely ate him. He was an australopithecine, literally a southern ape-man, a youngish specimen who, among all the scraps we have unearthed of his fellows shows one unique feature: he was killed by a leopard and probably carried, limp and bloody, to the cave where he was devoured. How can we be sure? Easy. The twin fang marks produced by the death-bite of the leopard are clearly shown through a section of skull. There is no sign of healing, so we know the bite was the cause of death. Perhaps what is even more darkly interesting is that, except for the fact that it's fossilized, that piece of skull could be a week old. Leopards, you see, haven't changed their style one bit in a million years. . . .

The question of why animals eat men has been one of more than casual interest to humanity as long as we have struggled to survive over those forces that threaten us, one of which is the man-eater of one form or another, even another man. Logically, a convenient point to start this discussion would be a consideration of the different broad classifications of man-eating and the animals who perform such unlikely (?) behavior.

A word of my proposed direction and the parameters of this book would also be in order at this time.

Man-eating is still a highly charged issue among people who have one or another emotional stake in wildlife behavior. For example, a scientist, naturalist or preservationist who has dedicated much of his life to the salvation of one or another of the great predatory species, perhaps the tiger or the leopard or, as is the current rage, the wolf, would be hard-put to admit, let alone develop the theme in any books he writes that his darling species has a greater or lesser tendency to eat people. Most books created on this defensive basis contain chapters that thoroughly attempt to discount any possibility of man-eating beyond, perhaps, the admission of one or two "rogues." Of course, this works both ways. Many of the great tiger, leopard and wolf hunters, to use the same examples, tend, possibly for reasons of moral justification, to try to point up the worst features of the species as a man-eater. Both writers are guilty of rampant humanization of animals, calling them on one side "wise" or "understanding," "affectionate" or "loyal" while the negative reviles them as "sneaky" or "bloodthirsty" or "cruel." None of these are, of course, true except in our imagined perception of human values in animals. Animals are animals and largely products of accumulated successful instinctual reaction.

All this emotionalism makes the researching of a book of this sort very tricky, as almost all sources can be suspect of shading their views. I have made every attempt not to fall into this bamboo-spiked potential pitfall. You will have to judge how fair or successful I have been.

Of the man-eating species, the seeming practical point of beginning would be the "natural" variety, of which there are not too many. Surely, both the Nile crocodile and the saltwater croc are just plain eaters of any protein they can catch under most conditions. So, in all likelihood is the red piranha and very possibly some of the sharks under some circumstances, although the shark is less certain. Very possibly the leopard is a "natural." The only way to delineate one form of behavior that leads to the same end as another style is consideration of *presumed* motive. That ain't always easy!

Next would be the "educated" man-eater, which is a classification of the higher animals only, traditionally cats, who have been taught by a man-eating mother to stalk, kill and eat people just as a "normal" offspring would be taught to kill antelopes and such by a "normal" lioness. This behavior is most common in lions and tigers; there is reason to believe that it may exist in the smaller leopard, but on a much lesser scale.

Desperation leading to hunger is probably the most common cause of man-eating, at least among the mammals (it certainly has created a lot of human cannibals!). When an animal is forced to a choice between starvation and gagging down a few evil-smelling people, he may well choose the latter course. The reason for an animal to become desperate enough to attack and eat man, which he or she may, from that point on adopt as a standard, may be any one of dozens. Among these would be injury to the predator which renders him incapable of hunting his normal prey either through a hunting accident or by incapacitating porcupine quills, broken teeth and the like. Age itself can be sufficiently debilitating to produce a man-eater through hunger desperation.

As recently as 60 years ago, many "experts" believed that no large predator mammal was a man-eater who was not insane, a mental deviate usually called a "rogue" in the case of buffaloes or elephants, (non-man-eaters) presumed to be literally a homicidal maniac if a carnivore. The accumulation of data has largely eliminated this concept, but there are still many cases of highly successful man-eaters who could handily be grouped under this category, and possibly, in some cases, quite correctly.

The last traditional grouping of man-eaters may be a large one: those who learned the vulnerability of man as prey by mistake. I have purposely left this a broad category, but a normal example might be for an African tribesman to stumble onto a pair of mating or feeding lions and be killed in a fit of temper. The first time, possibly nothing would happen, the body being abandoned. The second time, the big cat might well realize how easy it was and, tasting blood, have a bite or two of flesh. From then on, a full-time man-eater could well develop; in fact, several high scoring tigers are known to have begun in this way.

The obvious abundance of sudden gluts of human bodies in areas of predator activity such as produced by wars, plague, slaving activities and simply poor funeral practices can all trigger waves of man-eating among a wide variety of species, and unquestionably has in the recent history of the subject.

A particularly fascinating concept of the relation between man and those animals that eat him seems to me to be appearing in a brand new category of man-eater that has emerged only slowly since the turn of the century, when we first started a worldwide park system, ostensibly for the preservation of animals and habitat for their survival and our pleasure and study. What the cumulative effect of this idea may have been is to effectively, generation by generation, erase the natural fear that man has spent his whole existence as a predator to create in the other predators, to abraid the respect that generations of boldness and weapons use has brought for us in terms of respect. On the basis that many people are not eaten by predators because these big, dangerous animals fear man, this edge may be disappearing as we create an artificial familiarity with our old enemies. The behavior of bears in Glacier National Park may give one pause as will the most cursory glance at the recent man-eating activity of lions in many African parks where, 50 years ago, a rifle shot could clear square miles of lions. Today, it draws them—and other predators—who have learned that a shot means food, perhaps a zebra slain by a ranger or an animal culled from a herd. Man-eating is on the increase wherever there is increased close contact between man and the animals capable of eating him. So far as I can tell, this will only grow worse.

Despite the seeming increase of categories of man-eating, the good old-fashioned variety is still the hands down favorite among the traditional man-eaters. In fact, the following report from the Columbia, South Carolina *STATE* by the Associated Press and sent to me through the courtesy of Lt. Col. W. P. Alston, who has sharper eyes than do I, is less than a month old as I type these lines. It is a typical, or, if you prefer, classical pattern of predation on man. It begins with the headline: LION KILLS, EATS CHILDREN AND MOTHER: and is dated May 4, 1981.

> YANDE, Zimbabwe (Associated Press)—A lion killed and ate three children and their mother, and officials said Sunday they feared it may return and strike again.
>
> Government wildlife officials said the lion disappeared into the bush after the attack Thursday in the Dande tribal reserve in northeastern Zimbabwe.
>
> "If we don't get him soon there will be more suffering here," Game Warden Oliver Coltman said. It was Zimbabwe's second attack in two days by a lion, but game wardens shot the first after it killed a peasant farmer.
>
> In Thursday's incident (April 30) the lion was killing chickens and ducks penned near the family's hut when three children—alone and thinking the animal was a hyena—shouted to scare it off, Coltman said. The lion rushed the hut, killed the children and ate them, he said.
>
> When their mother returned later that night the lion, still feeding in the hut, attacked her.
>
> "It killed and ate her as well, then left," Coltman said.

Ah, well, I suppose some things never change.

Leon Parson

CHAPTER ONE

SHARKS

The water was as clear as a Parson's conscience, that incredible, pale jade hue almost matching thick sea ice, yet enveloping me in a warm-blooded embrace that must have been near 90 degrees F. I lazily turned on my back and looked over the black rubber flippers, scraped scars marked the hull of the tiny sailboat above; coral keloids from a dozen bumps and nuzzles against the second largest barrier reef in the world beside Australia's: what in those days was British Honduras and is now to be found in the sociopolitical section of the western Caribbean under "B" for Belize. That the country and the capital city had the same name, seemed to confuse and vaguely irritate nobody but me.

I was free diving with just snorkel, mask and flippers from a rented native fishing sailboat whose charter fee included the owner and captain, a wiry copperhided local Carib who, if not over-quick, certainly knew his way around the many islands of the group which included our base, Laughing Bird Key. Who, after all, could sail past a place with such an enchanted name without staying to share a night or two on the beach with the big, tropic moon, watching the twinkling carpet of feeding bonefish disturbing the phosphorus with tiny flak bursts of light on the flats? And who could ignore the shades of the old Mayan sailors who had left the shards of their broken pots all along the beach where they still studded the sand in some ancient form of punctuation? And, in the daytime, when the heat built up, there was always the weirdly beautiful hunting ground of gorgonias and coral grottos just outside the lagoon, where we would sail and Charlie—as my one-man crew was named—would putter around in the boat while I took the little single rubber arbalete speargun and prowled the watery jungle for a suitable overweight and hopefully unlucky Spanish hogfish, grouper or snapper. Such was my culinary mission that afternoon; an afternoon I will never forget under any circumstances short of being on the receiving end of a pre-frontal lobotomy.

I had spotted the grouper from the surface, his head sticking out of a coral niche twenty feet below. Flipping off the safety of the light speargun, I hyperventilated to saturate my lungs with oxygen and, watching the six-pounder through the face mask, dove on him with the speargun pushed out ahead, held by the pistol grip. From only a few feet, it was a cinch to nail him through the head, the spear darting off with the odd, slapping sound so peculiar to the underwater release of a speargun. The point of the spear drove through the fish somewhere behind the brain, the mortally wounded grouper going through the always amazing display of chameleon color changes as he fought the shaft and the length of nylon cord attaching the projectile to the gun, I guess about 10 or 12 feet of it. I had just begun to pull the spear back toward me by this cord when it happened.

It was 15 years ago, as I write this, and I can still tell you the sequence of every split second down to the last, crashing heartbeat that sounded like cannon fire in my pulsing ears.

It was a feeling; at first, almost the sort of sensation one gets upon realizing that he's left his wallet back on some store counter, a low voltage jolt of apprehension. Yet, it was more; not just apprehension, but also a mixture of that indescribable sensation of *realization* that you're being watched. Whatever it was, caused by some hidden mental sense or an unrealized movement of the water around me, I turned back over my right shoulder and just about collapsed with terror. Not a body length away and moving with irresistable power and surprising speed was the world's biggest hammerhead shark. The bloody thing looked like a submarine with the strange, flattened and extended lobe head appearing to be some new type of diving plane. At the end of the nearest projection, his left, an eye the size of an agate marble looked right through me. The mouth, I remember clearly noticing, was only partially open, just enough to see a hint of teeth rather than individual tooth points. There wasn't enough time to consider that exactly encouraging.

The initial shock of that godawful beast bearing down on me seemed to last a very short time before turning into pure and abject terror. Perhaps it's a very personal thing, but the idea of being attacked by something in the medium of water is identical to the common nightmare sensation of some irresistible horror overhauling you while you cannot run. Of course, under water, it's not a dream but fact. Man

(4)

can't move quickly in the dense medium of water, no matter what's after him, which, at least in my case, adds greatly to the sheer fear of the experience. An instant later, as I reflexively tried to pull my legs back, forming a sort of ball, there was a hard thump against my right thigh which knocked me swirling in the water, forcing quite a bit of air out of my lungs. I was momentarily aware of the shark streaming what seemed endlessly by and the incredible powerful currents washing over me that his movement produced. A couple of yards away, he was clamped onto that speared grouper and off, the speargun trailing behind him practically straight back. At this point, I suspect you might have been able to buy a slightly used but not abused set of snorkeling equipment at a most attractive price.

I didn't know what to do at this juncture. Somehow (and one would guess it the last thing I would permit to happen) I had completely lost sight of the shark, frantically turning around for a glimpse of his pearl-gray outline. I was afraid to try to make it up to the sailboat, about 30 yards off, for fear he'd pick me off half-way up. On the other hand, without any breathing apparatus and not much air left in my chest after the scare, if I stayed here I was for a certainty going to drown. The hell with it, I decided, springing off the bottom, secretly pleased that I had enough control not to cause a fuss or disturbance which might prompt the reappearance of you-know-who, as well as releasing air as I rose to avoid embolism or bends. I broke the surface what seemed a very long way from that sailboat, breast-stroking to cause the least possible noise. A couple of times I slowed down to look below with the face mask, but could see nothing of the hammerhead. At about five yards from the boat, I could stand no more. Making a break for it, I grabbed the top of the tiller and threw myself inside so fast it scared hell out of Charlie. Well, that was two of us!

Telling him what happened, I dug out my .357 Magnum revolver from my kit bag, airline companies in those days not yet having lost their sense of humor over the international transport of personal artillery. In the calm of that afternoon and the relative shallow, it wasn't hard to spot the shark as he came cruising back a few minutes later. Of the speargun, there was no sign, but I doubt that another hammerhead would have been in the vicinity. That this one paid some attention to the boat also pointed out that he was probably the same one, although he had lost about 40 feet in length from my first impression and was now down to about 12 feet long, which is still a hell of a lot of man-eating shark although hammerheads do get bigger than this one. To see what he'd do and since we didn't have a rod that would handle him, I got a most reluctant Charlie to throw him some cut-fish bait he had on board and the big shark took it readily after a cautious circle or two. As he swung around the boat, his high dorsal out and looking like a gray, leather sail on a toy boat, I shot him twice through that fin, doing him no harm at all. The other four rounds I had on board for the revolver I pegged at his head but, possibly because the

(5)

bullets were not jacketed or the angle was too shallow for penetration, they appeared to do no damage at all. As we ran out of bait, he ran out of interest and left. So, if you've ever happened to catch about a 12-foot hammerhead with a pair of bullet scars in the dorsal, one directly over the other and about two inches apart, there's a good chance I'm the guy who put them there.

That incident was one of my earliest run-ins with the "Tiger of the Sea," but it was, alas, not destined to be the last. One thing's for sure though, the experience, or at least the elements thereof, constituted probably the most common circumstances of shark encounter which, by definition, includes shark attack. I, for one, find it genuinely difficult to believe that there are living human beings who wander about the shallows of the world's littoral belt with strings of speared fish attached to their persons, oozing blood, lymph and whatever else a fish with a large spear hole tends to leak. The effect, if you've ever "chummed" for sharks in deep water with a mixture of blood, offal, ground fish and God alone knows what by ladeling the indescribable mess downtide until a tempted shark (or bluefish or tuna or whatever the species may be that is so enticed by the formula of the chum) comes along and takes the bait, a larger, presumably more attractive chunk of something resembling a successful Charlie the Tuna, is to make the person towing the string of dead and dying fish the sponsor of the chum "slick." That's something you don't want to be, as many hundreds of late spear fishermen would tell you were they able. In my case, I really wasn't doing anything more stupid than diving alone, not keeping my eyes open and causing, through the spearing of the grouper, a vast emission of what any self-respecting shark would consider an irresistable Indian love call, an aquatic dinner bell. These sounds are normally referred to, in the scientific sublanguage used by those who study sharks and kindred forms of critters, as low frequency pulsed sounds. Artificially reproduced between 20 and 60 cycles per second to imitate the sounds of a fish in distress, such as my grouper was emitting, they will draw sharks from considerable distances.

My problem that pristine afternoon was interesting; not only as it was transpiring but in afterthought. The grouper was the first fish I had tried to spear, so there was no offal or blood in the water that I was aware of. The hammerhead, which may have been a member of any one of the species *Sphryna* with the exception of the bonnet shark *(S. tibro)* which only reaches a recorded length of 3 feet, 7 inches; I wasn't watching that closely. Whether he was a *diplana* which reaches about 8 feet, in retrospect rules him out. Secretly, I have always believed him to be a *Sphyrna tudes,* the great hammerhead who reaches better than 15 feet, which is quite a lot of shark, especially if you're sharing the same grouper with him. Of course, he may have been a mere *S. zygaena,* who only grows to about 13 feet; quite sufficient to do in this writer, but not nearly so glamorous as to have been grazed by the presumably dreaded *greater* hammerhead.

I'd love to know how many hours I've spent thinking about that brush with eternity. Certainly enough to justify this chapter. As I see it, one of several possible scenarios may have been the case that hot, Caribbean afternoon. First off, that shark may have been shadowing thine truly with some basic recipes running through his rather limited Y-shaped brain for "Idiot Tartare" when I so rudely skewered that grouper and changed his menu. Or, he might just have been nearby, somehow eluding my hawk-like gaze until he caught the dinner call and went straight for the unfortunate grouper. I don't know. I don't *want* to know, either. I just write this stuff. Conclusions are up to you.

About all I can tell you is that a huge percentage of dead persons, which sounds nicer and far more personal than "deceased people" or "shark victims" were involved with the close proximity of fish they had speared or otherwise violated the civil rights of. The record books in fact, are absolutely obese with the similar circumstance of a fisherman or spearfisherman wading or swimming with his catch attached. Sharks are inclined to take either the man, the fish string or, in quite usual cases, portions of both.

There is one other aspect of the venture that doesn't jive with what all the best-selling experts say. Sharks *do* have extremely rough skins, covered with denticles so raspy they used to be used for sandpaper. As sword handles, they were also great coverings as they didn't get slippery with (presumably your enemy's) blood. Whenever this shark hide comes in contact with human skin, as in the way I was brushed, it's supposed to tear the skin off like a wood rasp. It didn't happen that way, and if you feel the hide of what you are absolutely double certain is a dead shark, you'll find the skin lovely and smooth *with the grain,* rough only against it. So, I ask, how does a shark tear off all those big patches of epidermis that even Oil of Olay couldn't fix up unless he swims backwards? I suspect a considerable amount of shark material is produced by barracudas who have never rubbed a shark the wrong way. . . .

The one positive point all shark researchers agree upon is that what's going on in a shark's skull is so completely unpredictable that the shark himself doesn't even know what he's going to do next. As the world's most widely distributed man-eater, found in over 70 percent of the earth's surface, water both salt and fresh, more and more accumulated knowledge clearly proves the point that probably no dangerous animal's habits (if, indeed, he has any reliable habits) are more inaccurately overgeneralized than those of the shark. Some people think we know quite a lot about the shark, but what we don't know is really astonishing!

We don't know how many species of sharks there even are, let alone the life cycles of even some of the best known, such as the great white. In fact, we can't agree even

(7)

what a shark is! Sure, if you'll sleep better, he *is* related to the skates and rays, who are all of the order called selachians. (Unless you'd rather call them Squaliformes, of course.) Sharks are also elasmobranchs, a subclass under the class *Chondricthyes,* which, when you remove the Greek means that sharks ain't got no bones at all. Their whole skeleton is cartilage. This, quite reasonably I think, causes many otherwise sane and lucid scientists to insist that sharks are not fishes at all but a completely separate life form. Considering the sharks' uniqueness, perhaps the eggheads are right. In any case, it's okay with me. I find them interesting, at least in a macabre light, because they kill and eat people all over the world in numbers we may not have even suspected. (The sharks, not the eggheads.)

In addition to a great deal of personal exposure picked up over years of living around sharks (as I type this, my office is 50 feet from the nearest land built on pilings over 12 feet of Florida bay water directly connected with the Gulf of Mexico. I saw a blacktip shark at lunch and inspected a dead 10-foot hammerhead yesterday afternoon.) I have researched the material for this chapter formally through at least a dozen books covering the known natural history of sharks, to tales of man-eaters, to novels, even getting into such highly technical papers as Thomas B. Thorson's *Osmoregulation in Fresh-Water Elasmobranchs* for a species I think deserves some special attention later on. Point is, I can't see this chapter attempting to be all things to all people interested in sharks when so much detailed information is available in your local library. Therefore, I'm simply going to tell you about some of the more colorful—perhaps an unfortunate choice of word—of the man-eaters as well as picking a few bones with the experts not borne out of my experience.

As seems consistent with the rest of what researchers have come up with about sharks, there is very little agreement as to which sharks are hard-core man-eaters and which are not. Some experts suggest that the species really dangerous to man of the roughly 350 we will generalize as representing the family, number about 40. Others claim much lower numbers, some no more than a dozen or 15. As records accumulate and are available through more central computer banking in shark research, it seems a fair observation that the higher figure are those at least pretty suspiciously implicated in human attacks. This trend is quite predictable when it's considered that even such great pioneers of oceanic science as William Beebe, who did so much great work in his bathysphere, firmly believed that there was no such thing as a man-eating shark! Public opinion, until about the First World War was, except for the lunatic fringe element of "old sailors," almost unanimous in poohpoohing sharks as a danger to man. They were to have their collective minds changed rather dramatically in, especially, 1916. The star of the controversy was an 8½-foot great white shark, the species made even more famous by Peter Benchley's

JAWS, which though a novel, was not without precident in real life along the coast of New Jersey. . . . But, first let's have a look, although brief, at the white shark.

A great white shark, *Carcharodon carcharias,* especially a large one, is just about the most impressive meat-eating animal we've had on Planet Ocean since the great dinosaur, tyrannosaurus, went west in the Cretaceous Period. Rather a chillingly interesting sidelight on the origins of Old Whitey, who at least these days is unquestionably an accomplished eater of man and anything else it would seem that he can get his jaws around, has to do with the unbelievable animal who—we very much hope—became extinct by developing into the modern great white shark. The measured record length for a white is 21 feet and better than 7,000 pounds. There are unrecorded but generally otherwise believable reports of whites up to 36 feet 6 inches long! But, this is almost literally small fry. The fossilized teeth of the ancestral *Carcharodon,* who was supposed to have gone out of business perhaps 20 million years ago, have been found in various places to be more than four inches long. This interpolates into a shark between 80 and 100 feet long at the minimum, weighing something like 50 tons! Sure. So he was a whopper. So what? He's been extinct for 20,000,000 years, right?

Um, er, well . . . It's this way: over the past few years more and more of these four-inch-plus teeth have been showing up, usually dredged from modern layers of ocean sediment into which they were not very deeply buried. They weren't fossilized, either. They were geologically fresh. Based upon that little piece of news, there may *actually be* a race of 100-foot long sharks somewhere down in the blackness. I wish I was kidding.

Like most well-distributed shark species, the white is locally known by a wide assortment of names. Most all tend to be darkly descriptive of this largest of flesh-eating sharks and include such tender titles as White Death, Man-eater, White Pointer, Blue Pointer and so on. Just where the "white" portion of the name comes in is questionable, as the *Carcharodon carcharias* isn't white except for his teeth and underside; the same as just about every other shark. His back, as a matter of fact, can be almost black. The outstanding features of the white are to be found in the nearly equal length of the lobes of the tail, similar to the mako (*Isurus oxyrinchus* or *I. glaucus*). The white is, however, *much* more ponderous than the mako and does not have such protruding teeth although they are much more equilaterally triangular and serrated like a steak knife. Guess why.

The great white is fortunately, a fairly low population animal, which can be illustrated by the figures from Florida that, of 100,000 sharks taken off that state's coast, great whites totaled only 27. Yet, they have a disconcerting habit of popping up when least expected just about anywhere and eating anything that's not tied to a treetop. From the best I can gather, whites must spend about half their time looking for food and the other half eating it. This has historically included a wide variety of humans (one of which may have been the basis for the Jonah tale. Remember, the Good Book refers to "a great fish" and not a whale.) from small boys to old ladies, dogs, pet seals, whole porpoises, sea lions, adult sea turtles, all types of fish including other sharks, pumpkins, goats, squid, entire horses and, in one outstanding incident, an African bull elephant!

Along the coast of Kenya, East Africa, in 1959, a lone bull elephant took it into his mind to swim offshore to an island for reasons best known to himself. About half-way there, a growing collection of very large sharks started to attack him and, as a feeding frenzy developed, literally tore the jumbo apart! From the description of eyewitness fishermen, at least some of the attacking sharks had to be whites from their size alone.

As we mentioned earlier, 1916 was a real vintage year for great white sharks along the New Jersey Coast. As a native-born New Jerseyian, I find it more than interesting to think of places I knew well and where I have myself swum being the center of a reign of terror and destruction in real life that actually was worse than the horrors dished up by Benchley's *JAWS*, be it book or movie. After all, an outbreak of serious man-eating by a wild creature only 15 miles from the city limits of New York must be considered a bit unusual, but, as is so often the case, truth can be stranger than fiction. . . .

It was the first day of July, a hot, sultry Saturday afternoon, little improved for 23-year-old Charles Van Sant by the crowded, muggy train ride from his home in Philadelphia, southeast to the summer resort of Beach Haven, New Jersey, at the southern tip of slender Long Beach Island. At last arriving at the accommodations Van Sant's father and two sisters shared, Charles merely fished his robe and two-piece swim suit from his suitcase and bolted for the beach, the cooling, swirling, surf soothing on his legs as he waded out a few yards and began to swim.

He was a strong young man with the muscular ease of the practiced swimmer. We don't know whether his mind was on the Great War in Europe or on the more than 100 people who had died so far in the epidemic of polio that ravaged New York City that season. Probably, neither the trenches nor the infantile paralysis were in Van Sant's thoughts; just the cool smooth sensation of the water flowing over his body as he pulled himself along with easy, powerful strokes, after 100 yards turning back. Certainly, he never saw the big, dark fin that reared like a giant scalpel blade behind him, slicing inexorably closer and closer through the calm surface of the Atlantic. On the beach, people noticed the sinister sickle and screamed unheard warnings to Charles, the black form bearing down on him with a terrible certainty as he blithely swam on, lost in his unknown thoughts as death silently edged closer. As it was nearly on Charles Van Sant, the people on the beach stopped their screaming and stood on the hot, sloping sand, watching with mute, fascinated shock. He was only 30 yards from the sand when the onlookers saw the dark, long form reach the swimming man and the surface foamed, a growing pinkish-red blossom of blood blooming all around.

Without a moment's hesitation, a bystander named Alexander Ott, himself an ex-olympian swimmer, ran for the water and dived in, streaking toward the struggle. Presuming that, as a witness, he was aware that a man-eating shark was eating his prey, to jump into that same water was an act of extraordinary bravery. When Ott reached the patch of gory water, he saw the shark feint at him and then flash away back seaward. Somehow, Ott managed to grab Van Sant and, with help from gagging bathers, dragged him up onto the beach. Most of the meat had been torn and eaten off both legs. Surprisingly, considering the extent of his wounds, Charles Van Sant did not die of shock and blood loss until that evening.

One of the most astonishing aspects of the killing and partial eating of Charles Van Sant 30 yards off the shore of a busy New Jersey beach was the reaction of the state and the nation to the incident.

There was *no* reaction. The whole episode was virtually ignored, not so much as a shark warning being spread to neighboring communities! It was well known that sharks just didn't eat people. And, if they did rarely dine on some poor chap twiddling his toes off Fiji or some other far-off place, they most certainly did not eat paying guests on the shores of New Jersey! After all, there had been an unclaimed reward of $500 put up by a New York banker payable to anybody who could conclusively prove that any shark had attacked any bather north of Cape Hatteras. It had gone uncollected for 30 years by 1916, and $500 was a huge amount of money in those pre-inflation days! True, the foot of a woman, still wearing a tan shoe and stocking, had been taken out of a cut-up shark three years before that had been caught at Spring Lake, almost 50 miles up the New Jersey coast. Even the skeptical "experts" acknowledged this case, but solved the problem by suggesting that the owner of the foot was probably a corpse before being eaten. It did not prove that a "cowardly" shark would ever attack a living swimmer.

It was the following Thursday, about 2:00 in the afternoon of July 6th, oddly enough at the same place where the mysterious foot had been recovered three years before. Spring Lake, closer to New York, was among the most fashionable of the resorts catering to the prominent from there as well as from Philadelphia. With the outbreak of infantile paralysis getting worse in New York, the seaside town was crowded for this early in July, and the bellhop of the Essex and Sussex Hotel thought it looked like a good summer for tips. Charles Bruder, 28, was practically a fixture at the place, having worked at one hotel or another since he was eight years old. He supported a mother in Switzerland as well as himself and, with that Thursday afternoon off, decided to go swimming as he often did.

On lifeguard duty that July 6th were two friends of Bruder, Christopher Anderson and George White, young men who knew that Bruder was an excellent swimmer. When he kept going past the lifelines, neither of the guards thought anything of it as they realized that Bruder often did this. As usual, the lifeguards continued scanning the few bathers, not noticing that Bruder, who had been swimming quietly along, was not in sight.

And, then, the screech of a startled woman lanced the hot sunshine, shouting something about a man in a red canoe turning the boat over. Anderson and White quickly looked to where she pointed and immediately realized that the red they saw was not the paint of a canoe. As they ran for their own rescue boat, the face and one arm rose from the bloody water, probably lifted out with the impact of another terrible bite. Getting close, the lifeguards got an oar to Bruder, which he was able to hang on to, and worked him up to the boat. In shock and agony, bleached from loss of blood, he gasped out that a shark had bitten his legs off, then fainted. By the time he—or what was left of him—was dragged up onto the beach, Charles Bruder was well and truly dead. The doctor who tried to save him began to work on the vacationers who puked and passed out at the first sight of half-eaten Bruder's body. Somebody finally covered it with a woman's coat.

If the attack on Charles Van Sant the previous Saturday in Beach Haven's surf was ignored as some kind of a freak accident, this was certainly not the case in the killing of Charles Bruder. All the New Jersey seashore went mad. The pronouncement of the well-known physician Dr. W. G. Schauffler, Surgeon General of the New Jersey National Guard, (who happened to be at the beach and inspected the body almost immediately after death) was that there was no question that Bruder had been killed by a shark. The body's right leg was horribly ripped and bitten off below the knee. The left foot and lower leg were also chewed off and there was no flesh on the bone left below the knee at all. The leg above the left knee was also bitten to the bone, complete with tooth-marks and a chunk of meat "as big as a man's fist" was missing from the right, lower abdomen. Well, after all, an animal that has been witnessed to bite a full-grown porpoise *in half* with one jagged razor snap wouldn't waste too much time on a human's legs!

Shark fever swept the Atlantic coast, huge sharks reported everywhere, gangs of armed men roaming the beaches, and the resort hotel owners going down with terminal migraine with the loss of over $250,000; occupancy was down 75 percent in some areas. Contractors were cleaning up installing "shark-proof" netting in bathing areas as the hysteria ran its course.

Of course, there were still the "experts" to come to the rescue to debunk Dr.

Schauffler's medical opinion. Dr. F. A. Lucas, Director of the prestigious American Museum of Natural History, flatly stated that no shark's jaws were strong enough to actually bite off a man's leg and cause the kind of mayhem Schauffler claimed. He was joined in the opinions of Dr. Nichols of the Department of Fishes of the museum as well as by "authorities" of the Brooklyn Museum, that sharks were basically not dangerous. Even the U.S. Commissioner of Fisheries, Hugh Smith, opined that the odds on another attack were miniscule. Of the $500 prize offered by the New York banker, I can find no additional information.

While the experts and the neophytes were arguing, an 8½-foot long eating machine was casually headed north in the quiet Atlantic. After a day and a night, Spring Lake lay well south, along with Belmar, Asbury Park, Long Branch and even Sandy Hook, where the shark turned west, taking his time to cross Raritan Bay, right between the Amboys and then slightly south into the mouth of Matawan Creek. Why enter tiny Matawan Creek, hardly more than ten yards wide on a good tide? I didn't say great white sharks are smart; I said they can swim anywhere they want to!

Precisely what date the shark covered the 25 miles between Spring Lake and Matawan Creek is conjecture, if in fact the shark that killed Van Sant was the same shark who killed Bruder was about to commit multiple murder again. For my money, it was. I can't believe the laws of probability would permit roughly a 70-mile long section of coastline, which had never in memory or even literary tradition had any kind of shark attack to suddenly lose two lives and, as we'll see, several more. That's just too much to believe.

We *do* know that one 14-year-old boy, Rensselaer Cartan—who was swimming with his pals at Wyckoff Dock on Matawan Creek, sometime between the 7th and 12th—half-fell, half-jumped from a piling and hit something coarse and rough enough to draw blood from abrasion from across the skin of his chest in what was normally open water. Frightened, he got out of the water and shouted to his companions to stay out because there was something unauthorized in it. But, the kids payed no attention and nobody was hurt; not that day, anyway.

It's sort of interesting that on the 11th of July, my mother's 12th birthday, a fisherman named Tarnower snagged and landed a 9-foot shark of unidentified species only 40 yards from the low water mark at Belford, precious near the mouth of Matawan Creek. It must have contained nothing by way of connection with the deaths as the incident was practically ignored by the press, who were most certainly not suppressing shark stories!

The next day, Wednesday, July 12th, 1916, a retired sailor and resident of Matawan was crossing a recently constructed drawbridge a mile and a half downstream of the rising tide that marked the worn, wooden pilings at Wyckoff Dock.

The sailor, named Captain Charles Cottrell, was stunned to look into the water of the innocuous-looking creek and clearly see the dark form of an economy-sized shark moving up the tide with good speed. Cottrell, who must have had plenty of experience recognizing sharks after a lifetime as a blue-water sailor, immediately shouted to a pair of workmen on the trolley bridge to look, and they also clearly saw the shape of the shark. So impressed were the two workmen that they ran like hell for a telephone to call the Town of Matawan's Finest, headed up by the proprietor of a tonsorial parlor, barber John Mulsonn. Cottrell, meanwhile, had hoofed it back into the town's small center to try to warn residents and others who, including Chief of Police Mulsonn, thought the idea of a shark two miles from the bay up the skinny water of Matawan Creek was barely short of hysterically funny.

Frustrated and probably a bit angry at his reception, Captain Cottrell continued along the street, especially trying (without success) to warn off groups of children on their way to swim at their customary spot off the Wyckoff Dock. One shop he did stop by to warn was that of the son of the retired Commodore Watson H. Fisher, who had made his fortune in shipping and had retired to Matawan where his hulking son, Stanley, had opened a dry cleaning store. As the endeavor was new and business slow, the well-liked Stanley had a sideline selling men's suits. For one of these, he had accepted payment in a paid-up $10,000 life insurance policy. It was crazy, his friends told him. At 24 and in the bloom of health, he would be the last person wanting a life insurance policy!

By 2:00 in the afternoon of that sweltering Wednesday, things had somewhat calmed down in Matawan, the citizens more interested in getting cooled off than in shark stories. The usual gang of naked boys was gathered at Wyckoff Dock, including Lester Stilwell, excused early from his normal chores at his father's place of work, Anderson's Saw Mill. With the skinny 14-year-old Lester was a relative of Rensselaer Cartan, the boy who had gotten the scrape from the unknown fish earlier, Johnson Cartan; the young Albert O'Hara and several other local boys. It was probably just short of 2:30 in the afternoon.

Albert O'Hara, one of the youngest at 11, turned before starting to climb out of the creek when Lester Stilwell shouted for the rest of the boys to watch him float. As Albert O'Hara turned, he felt a rough blow as something struck his right leg a hard but glancing punch. Looking down, he was petrified to see a huge form bolt past his legs and torpedo straight at the clowning Lester Stilwell. Another lad, Charles Van Brunt, also saw the terrifying shape lunge at Stilwell, impressed with how completely black it seemed. As it struck Stilwell, there was a flash of snowy belly and gleaming white teeth closing over the slender form of the boy. Both Albert O'Hara and Charles Van Brunt knew they had seen a killer shark. As the horrible cloud of red blood and torn meat stained Matawan Creek like an eerie, silent, deadly flower, there was numbed astonishment—then pandemonium.

As soaking, naked boys fanned out from the scene at Wyckoff Dock, some heading directly into town and others seeking adult help wherever they could find it, the report—as might have been expected—became garbled. Most people only knew that a boy was in trouble at the creek, but whether the cause was a shark, as reported by Captain Cottrell earlier or because the boy was subject to fits, was immaterial to big Stanley Fisher. He grabbed his swim suit, turned the shop over to his errand boy and headed for the dock. Either upon arrival there or on the way, he became convinced that the Stilwell boy had been taken by a shark. Fisher knew the bottom near the dock, where the channel was only about 20 feet wide, hardly the place to expect a man-eating shark, but became convinced that the shark was lying in a deep hole across the creek from the pilings with the boy's body.

Taking charge, Fisher ordered the creek blocked with weighted chicken wire so the shark could be forced into shallower water, but there wasn't enough wire. Nonetheless, although there were already several people "blind" diving for the body, Stanley Fisher also entered the water, swam over the deep spot and dove into the dark depths. He shouldn't have done that. . . .

He was next seen, moments later, by the man who helped to pull him from the creek, Monmouth County Deputy Detective, Arthur S. Van Buskirk, who was in the bow of a small boat when Fisher appeared in the middle of another bloody flurry of water near the far shore. Van Buskirk got the man in the stern of the boat to start the engine and headed for Fisher, who was in waist deep water, both hands clutching at what was left of his right leg. Facing away from the dock, spectators couldn't see the extent of his injuries, but Van Buskirk could when he tried to lift the 210-pounder and was only able to drag him halfway out of the water. When the boat swung back toward the dock, onlookers fainted. The flesh was completely stripped from the right thigh, dark arterial blood pumping from the terrible bite. Although still conscious, he tried to speak but could not.

Stanley Fisher was carried as gently as possible to the Matawan railway station about a quarter mile away after being tourniqueted. By the time he reached the station, a doctor had been found to treat him, but there was nothing to do but improve the tourniquet. After nearly three hours in the sweltering heat, the train came through and picked up Fisher. He died in the operating room at Monmouth Memorial Hospital. Before he succumbed, he managed to murmur that he had found the body of Lester Stilwell and had taken it from the shark. Lord, what a price to pay for a cadaver.

Shortly after the then-still-alive Fisher was dragged from the creek and carried to the station, the crowd returned to town for a load of dynamite to pulverize the shark

(15)

they thought was still in the 20-feet-wide section of the creek. The charge was rigged, but just before being blown, a boat hove into view driven by a local lawyer and carrying a 14-year-old boy with a badly mauled leg, wrapped in bandages. He was hustled off to St. Peter's Hospital in New Brunswick (why, one idly speculates, was this not done with Fisher rather than letting him sit at the railroad station for almost three hours).

Joseph Dunn, as the mutilated youngster proved to be, had been swimming with friends and his older brother a half-mile down Matawan Creek when word of the attack on Lester Stilwell arrived. Immediately, the boys cleared out of the water, but Joe hadn't been quite quick enough. He was actually climbing the ladder at the dock of the New Jersey Clay Company wharf when the (or *a*) shark engulfed his right leg almost completely. Joseph said later that it felt like being chewed by a big pair of scissors as his leg went down the shark's gullet.

As Joseph shrieked that a shark had him, his brother and two other boys gripped him and hung on, the shark trying to tear Joseph free. After several moments of the awful tug-of-war, with the life of Joe Dunn the prize, the shark released him and glided away. In less than an hour he had struck three times, killing twice, and the question now was whether the score would be three. Despite the badly slashed muscles, veins and tendons, Dr. R. J. Faulkingham not only saved Joe, but a day less than two months after the attack the boy walked out of St. Pete's under his own steam.

Rather understandably, the Matawan area went crazy in a mass attempt to kill the shark, still presumably in the creek although obviously on his way out to sea as he had attacked Joseph Dunn along the way. Whatever the case, a full-blown shark hunt, featuring everything from explosive charges to old-time whaling harpoons, swept Matawan. The only thing as frequent as the explosions were the imagined reports of sharks who just managed to get away. Finally, things died down and Matawan buried its two dead. I have been unable to determine the exact damage to Lester Stilwell's body except to mention that it was reported bitten across the middle by witnesses at the first lunge of the attack. That it was recovered at all is proof positive that Stanley Fisher had done as he had said with his last dying, agonized breath: rescued the torn body of Lester Stilwell.

On the 18th of July, nearly a week after the triple attack, none other than the original sighter of the shark, Captain Cottrell, helped by his son-in-law, managed to net a 7-foot, 230-pound shark of unrecorded species, certainly containing no odds or ends bitten from the bodies of either Lester Stilwell or Stanley Fisher, or they would have been prominently mentioned. Still, there were no barnacles growing on old Cottrell, who displayed the iced leviathan, the "Terror of Matawan Creek" at a dime a crack as long as ice and spectators lasted.

1916 was not exactly yesterday, and it's difficult over the hazy perspective of the years to figure out how Cottrell had been able, six days after the attacks, to claim that *his* shark was the killer. You see, *two* days after the bloody going-on, the animal that was almost positively *the* "Terror" had already been killed, also caught in a net by one of the vast posse of prospective shark-slayers in Raritan Bay. It was an 8½-footer which apparently went unweighed but whose stomach contents were definitely identified by the same Dr. James T. Lucas, Director of the Museum of Natural History, as well as Dr. Nichols, who had also positively denied the existence of man-eating sharks, as unquestionably human. Among the grisly catalogue of selachian heartburn were 15 pounds of human flesh including an 11-inch section of a boy's shinbone as well as a generous portion of human rib.

The shark was also confirmed to be a great white, and the circumstances pointed strongly toward the conclusion that this single shark was responsible for all five assaults. That's pretty good work by all those "experts!"

Because of the incident, America did a one-eighty as far as sharks were concerned. A congressman tried to have a bill passed raising money to fight the problem. Guess which state he was from! In general, though, 1916 seems to have been a freak year for the frequency of sharks, especially on the eastern seaboard, as was confirmed by several authorities. Of course man-eating sharks did not cease their charming antics with 1916; rather, people became more aware of them and took more precautions. With the exception of going to the mountains rather than the beach, though, there was bloody little that could be done. Oddly, another rash of attacks by great whites took place in 1960 off New Jersey, but, after all, I think we've given the devil his due as far as *Carcharodon carcharias* is concerned so let's leave Whitey for a while and take a look at his partners in crime.

*   *   *

Although, at least in the opinion of the author, the great white would be my last choice of species with which to share accommodations of any sort, this shark probably kills less people than others. The reason for this is that he's so relatively rare compared to hammerheads, blues, lemons, makos, tigers, white-tips, black-tips, porbeagles; the list of proven and implicated sharks goes on and on. Yet, the species which very possibly is the overall high scorer is one of the least known and for sure, the least understood. In the phone book, you'd probably find him under *Carcharhinus leucas,* but elsewhere around the world, he's called the cub shark, ground shark, requiem shark and bull shark. Since I know him best as the bull, let's stick with that.

As long as we're on this name business, this would be the time to point out that *C. leucas* is rather unusual: he's anadromous, which means that he can and does run back and forth between fresh and salt water any time he damn well feels like it. Until recently, this trick led to his additionally being known by the location where he was found, to wit: Lake Nicaragua shark, Gangese shark, Zambezi shark (with corresponding Latin monickers). Today, after a good deal of research, the Ichthyological Powers That Be have come to the same conclusion that I reached (albeit without the validity of their empirical process) back in the late 1960's after spending a great deal of time in Nicaragua, both on the largest lake and its 130-mile long drainage to the Caribbean, the San Juan River. When I finally saw the Zambezi River shark, it was like bumping into an old, well, I almost said "friend." What I meant was that the sensation was similar to a recurrent nightmare. The point I'm getting at is that the bull shark will cheerfully bite the sweet, baby Jesus out of you whether he catches you in salt water, brackish bays or estuaries or the sweetest fresh water you can find, from Nicaragua, Guatamala, Louisiana or even 50 miles up the Hudson from New York City, to southeast Africa, New Guinea, India, Malaya or Iran. Since far and away the great majority of sharks who bite chunks out of people are neither caught nor identified, it's hard to hang labels accurately on species. But, not so the bull shark. It has been determined that the bull is the same shark in fresh water rivers and lakes in so many parts of the world, with a couple of possible exceptions; a shark was caught 2,300 miles upstream from the mouth of the Amazon, at Iquitos, Peru, which was unidentified as to species (what the hell, who would know a *Carcharhinus leucas* from an oversexed woodchuck in Iquitos, Peru, on the wrong side of the Andes and 2,300 miles from a practical access to salt water) I'll put money that it *was* a bull shark. It is my suspicion, and taken only after at least 15 years of on-the-spot, full-time wondering, that more people are injured, killed, eaten or just plain disappear because of bull sharks than any other species, barring those years when there is a large sea disaster and mass feeding patterns on survivors by pelagic (deep water, open ocean) sharks untypical of bulls occur, which throws the figures awry.

Again, the reliability of "official" statistics on man-eating of any kind, either by mammals, fish or reptiles is by definition unreliable. If somebody's *eaten,* mate, he's bloody well *eaten*! When you're eaten, you just *aren't* any more. Unless witnessed, people tend to be marked up as "missing," particularly in the matter of aquatic attack such as by shark, crocodile, piranha, etc.

It has always struck me as rather odd that my career would have brought me into such close proximity with the bull shark, both in Nicaragua where I was financially involved with the only sportfishing camp on the San Juan River as well as during that long period as a professional hunter in southeast Africa, where the same species, at that time known and segregated from the identical Lake Nicaragua shark as *Carcharhinus* (sometimes *Carcharinus*) *zambezensis,* was considered absolutely ferocious and by far the greatest man-eater in the area of the Zambezi mouth as well as in Natal, South Africa. This shark has been known to travel as far as 120 miles up the Zambezi and, although not usually advertised, has attacked humans as far as *150* miles upstream from the mouth of the nearby Limpopo, the border river between South Africa and Mozambique!

Okay, to speak of a rash of attacks by great white sharks off New Jersey or Australia is to speak of four or five deaths being a real state of mayhem. Yet, bull sharks killed or mutilated 27 persons as far as 90 miles from salt water in the Karun River of Iran in just one seven year period in the '40s. One of these was a British soldier, an ambulance driver who was washing his vehicle in ankle-deep water when severely mauled at Ahwaz. In Nicaragua, just along the San Juan River, I actually knew one man killed when his *cayuca* dugout turned over, eaten by sharks, and the father of one of our employees, Emilio, was eaten about five years before my arrival in 1965. According to the local people, at least when I knew them before the devastating earthquake and possibly even more devastating political upheaval after that, at least one person was taken in the river each year, and about three in Lake Nicaragua during the same period.

If you ever want to wander around a weird piece of geography, try Nicaragua for starters. Mix in the strange geology, the odd facts and guesses associated with the lakes and main river, add a slew of salt water fish who aren't supposed to be there and you've got a several thousand square mile puzzle that's a real challenge so long as you don't try to swim in it. That's not a challenge; that's suicide. Long as we're on the subject, let me tell you about it.

The thinking about the aquatic geology of the Nicaraguan region has changed— to quite a bit more reasonable stance, as far as I'm concerned—since the late 1960's. Previously, the authors of shark books obviously had not spent time on the lake or its drainage to the sea, as nearly universally it is claimed as semiunique for the fact that it contains *four* species of otherwise salt water fish. Right off, this is wrong, although the four species are among others aggregating a greater total than four. Normally, the so-called "fresh water shark" (previously *Carcharhinus nicaraguensis,* now agreed universally to be *C. leucas,* the bull shark, the ubiquitous man-eater) was at the head of the list followed by two brands of sawfish, which are closely related to the sharks, skates and rays, and also the tarpon, that great sport fish that brought me to the San Juan in the first place. What never seems to be mentioned is that both the snook, locally called *robalo* in Spanish *(Centropomus undecimalis)* and the jack crevalle are also present, especially the former in great runs up the San Juan during proper moon and water conditions.

Perhaps much of the controversy (and a chapter of Lineaweaver and Backus's book, *THE NATURAL HISTORY OF SHARKS,*—Lippincott, New York, 1969,

1970—dealing with this matter is called "A Matter of Controversy") is due to the fact that quite a number of fish not realized to be so are anadromous, able to travel and exist between both fresh and salt water. The obvious are the salmons, shad, sturgeons and more traditional types, but there are many others including jacks, snook, trouts, scats, (especially *Scatophagus argus*) eels, sticklebacks, guppies, chubs, archerfish and even the largemouth bass which I have personally caught in pure salt water in Florida. Undoubtedly there are other examples that don't readily come to mind, but I think you get the drift. It's upstream. . . .

Nicaragua has been doing a geologic hoochie-koochie for the last few hundred thousand years and, from the last extraordinary tectonic holocaust, there's nothing new on the horizon except the imminent prospect of a mountain or two that wasn't there yesterday. Managua, at least on a probability table (if it's still there when this book reaches publication) can almost be counted on to disappear at least once per century. It was leveled last in 1931 before the current debacle and put on its best show in 1835 when a volcano named Coseguina did a triple flying somersault in pike position which made Mt. St. Helens look like a leaky carbuncle. The boom stopped daily activities in Bogota, Colombia, 1,100 miles away and pieces of the cone the size of small real estate developments landed as far as 150 miles out to sea! Against a background like this, featuring 23—count them, 23—live volcanoes it's easy to see where the controversy arises about the immensely exotic selection of aquatic creatures that eat Nicaraguans and willing foreigners.

The original published account of the toothy denizens of Lake Nicaragua and the San Juan River was from the facile pen of the Spaniard, Oviedo y Valdez in 1526. His remarks in *OF THE NATURAL HISTORY OF THE INDIES* were well considered and confirmed by the American *charge d'affaires* in Central America 326 years later Mr. E. G. Squier, who quite agreed with Oviedo y Valdez in that what would later be decided to be bull sharks swarmed the lake and were called locally *tigrones*. (I speak Spanish but this word must be a corruption for a reference to either "tiger" or the local term for jaguar which is *tigre*. *Tigron* does not appear, either, in Cassell's *NEW COMPACT SPANISH DICTIONARY.* What the hell, you wouldn't have to speak Spanish to know that both writers were speaking of a shark that bit a carload of people and was not noted for its courtesy to women and children. There are times when one does not need the emergency services of Berlitz!

The central channel of the San Juan has been subject to a great deal of hydrological change over recorded history, yet even the most perfunctory examination of the river would yield the fact that the "Lake" Nicaragua sharks are certainly not landlocked. The original theory of the origin of the lake and the presence of the supposedly salt water species was that, in some dim age, both Lake Nicaragua and the smaller, adjoining Lake Managua were bays of the Pacific Ocean. With one of the eruptions or perhaps because of an earthquake, a barrier of dry land rose, sealing off the bay from the sea and forming the lakes. Of course, says the theory, all the natural salt water species were trapped in the new body of water which became ever more brackish through rainwater dilution until bursting out through the San Juan River to the Caribbean and turning into a completely freshwater lake. Those tough species such as the sharks, sawfish, and tarpon (with their rudimentary lung which permits partial air breathing), were able to withstand the change and so survived.

When I first heard the theory, I accepted it without much thought. Involved in the area, especially the river, as a fishing camp operation offering first-class tarpon light-tackle angling around the El Toro Rapids about 80 miles upstream from the sea, the origins of the fish didn't matter a great deal to my wallet. Yet, the more time I spent in Nicaragua, the less I had heard as gospel seemed true.

For starters, if the lakes had been a Pacific bay, what were tarpon doing in it? The Latin designation of the tarpon is *Tarpon atlanticus* and it doesn't occur in the Pacific, at least nowhere near Nicaragua. Oh, well, I figured, they could have come up the river in a geologically later time slot for whatever reason turned them on at that time.

I had only visited and fished the river one day before I realized that if Lake Nicaragua had a reputation for scads of the blighters, the numbers of sharks in the San Juan must be terrific. Incidentally, getting there was all the fun. One hired the necessary numbers of light Cessnas—172s, if memory serves—and the Nicaragua Air Force would fly you across the 125 miles of the lake to the jungled strip at San Carlos, a grubby lakeside town at the juncture of the lake with the San Juan having an especially bad reputation for man-eating sharks. I remember the flight especially well as, on this instance, arriving just at dark, the young lieutenant piloting me and most of my party's gear overshot the dim runway, tore a wing completely off and scared about 30 years off my life. That we did not burn has remained one of the greater mysteries of my short tenure on this rather baffling planet.

The history of the San Juan, if you care to skim it quickly with me, is mighty interesting. Today, because of landslides and other phenomena associated with earthquakes and other related forms of geological indigestion, it is not as navigable as it once was. But, back in the gold rush days, before the Panama Canal, it was possible and common to take a ship to the mouth of the San Juan (the river itself was deep enough for ships in recorded history) and then switch to riverboats up the stream and across the lake to waiting stagecoaches who carried the gold-lusting miners to other ships on the Pacific side and north to the gold fields. That it was a good, sound idea that worked well is attested to the fact that much of the operation was run by Commodore Cornelius Vanderbilt. As recently as 1882, the steamship *Victoria* out of Wilmington, Delaware sailed completely up the river and into the lake. Why anybody would want to do that then—or today for that matter—is slightly beyond my ken of commerce.

I don't want to get the powder before the ball here, but there is a great deal made locally about whether or not there are *two* types of sharks, as claimed by the natives. These are supposed to be the reddish-pink bellied *tintorero,* bigger than the white-bellied *visitante* or *immigrante* who is supposed to be a sea-run shark of the same species who, for some unknown reason has decided to give up all that lush, high cholesterol living in the Caribbean and run up the river. With very little else to talk about if you happen to live on the shores of either Lake Nicaragua or the San Juan River, I can see why such a matter would be of equal import as the amount of rain, the question of how far away one can hear the big, hairy brown banana spiders stamping their feet on the tin rooves of the camp *cabinas* or whether the tarpon are biting. Conversation is limited somewhat by the fact that the tarpon are just about always biting.

Now, in those days as well as now, about the only thing I reckon could be more exciting than constantly fighting big tarpon on ever lighter fly tackle was the temptation of all those sharks. Man, they were all over the place, even attacking hooked tarpon and either forcing me to break the silver flyers off on purpose or see them eviscerated in a big wallow of red water as one, two or more of the sharks struck the hooked fish. Basically, it was exciting as all dammit, and I decided that it was high time to shift the emphasis to thinning out a few sharks when it became clear that even those tarpon who successfully evaded the shark attacks while hooked were being slaughtered when they slid back into the current, exhausted after the fight.

Back at camp, digging through some of the mouldering books of fishing kept for the guests, I found an old copy of the U.S. Navy's "Shark Danger" Listings. The idea of assaulting some of those loitering elasmobranchs seemed a bit less attractive when I read that the Lake Nicaragua shark is rated as a dead equal with ferocious tiger shark at a $+2$ listing. Still, calling in my guide, Jose, I asked him the best way to even things up for the tarpon. *"Arpón,"* he said without hesitation. "Harpoon."

That same afternoon, Jose was dodging the rapids rocks in the big freight canoe we used to get to the airstrip or rarely downstream to the village of El Castillo (The Fortress). This was a mighty interesting spot with an imposing stone fortress towering over a shanty town of huts and shacks, one of which was a general store. Here, we were able to buy two hand-forged harpoon heads with skirt-like bases and a long, barbed spike which fitted over the point of the homemade shaft. Jose, who spoke English like I do Cretan Linear "B," and I got along in my back alley Spanish which, since at this time there were no American clients in camp, was improving through "total immersion." This made it hard to get a deep breath, but did wonders for my rolling "RRRs." I never did get the difference straight between the Spanish for "lettuce" and "owl," but we managed to buy the spearheads okay. While there, I also bought an American Indian head penny found at the fortress, lost either by a forty-niner or perhaps one of the U.S. Marines during one of the Nicaraguan campaigns. It lasted three days before I lost it, anyway.

While Jose was off fraternizing and I was crawling around the battlments of the castle, overlooking a mean set of rapids, I ran into a *Nicaraguense* who had been to sea and spoke excellent English. As soon as the subject turned to sharks, he told me that there was one especially big one who hung around the top edge of the rapids and had attacked several canoes. About six months earlier, this shark had managed to upset one *cayuca,* in plain view of the village. The occupant, a man alone, had been grabbed by the leg and had lost huge pieces of meat and muscle before the current drifted him against a rock which he was able to climb to stay out of reach of the big shark. By the time he was rescued, he had bled to death although his hands could be seen tightly gripping the leg above the wounds in an effort to staunch the flow of blood.

That particular shark, at least as the man at the castle told me, was famous as *"El Amarillo,"* the "Yellow One" for what was said to be the odd cast of color of his skin, which was supposed to be a dull, orange-brown. I'd love to tell you that I saw him, 11 feet of badly-bleached canary death cruising around the head of those rapids, but, although plenty of sharks could be seen from the heights of the castle, none resembled that one. Jose, from the best I could understand, pantomimed out that *El Amarillo* was no local version of the bogey man and was one shark to be avoided. He had seen it several times and a friend had shot at it. The current concensus was the collective hope of the village of El Castillo that the bullet had been effective although no body had been found. Maybe the others had eaten it? *¿Quien sabe?*

Twice, on the way back upriver, Jose nuzzled the nose of the canoe to shore and, with precise swings of his machete cut harpoon shafts of jungle hardwood, leaving them half-trimmed to fit that evening. That night, dinner took forever, even though it was superb snook filets, and I went to bed early, thinking about our harpooning expedition the next day. The last thing I had asked Jose, who had been born on the San Juan, was what he thought of the whole idea. *"Divertimiento, Patron,"* he grinned wryly, *"pero bien peligroso."* "Interesting fun," he meant, "but plenty dangerous." Well, we'd see.

The sun was still low and the river air moist enough to bead the spider webs with pearls as Jose's mahogany knuckles tightened and showed white as the long anchor line came taut. In a practiced motion, he deftly dropped a loop over the bow post and the dugout swung to a halt.

We were anchored in "the pool" as the natives know it, where the main current channel of the El Toro caresses a shallow bar which separates the fast water from the deep quiet of a huge eddy. At the moment, at least 30 big slate-colored fins cut the surface; slicing the patchwork of tropical sunlight filtering through the treetops overhanging the river to the east. Antenna-tailed parrots and colorful dragonflies darted about overhead like nervous spectators to the coming encounter. Even the air was oppressive and ominous. I felt as if they were about to open the doors to the Roman Arena with me as the star attraction, and I had left my sword at home.

As we watched, every few minutes a big shark or two would splash across the shallow bar and slide into the deep water hunting grounds upstream.

Watching Jose study the menacing fins, I sensed that he knew the ferocity of the beasts beneath the surface of the chocolate colored water better than an ordinary fisherman. Needless to say, I was happy as hell to have him along. Emilio, the guide who had lost his father to the sharks in about 1961, had turned me down flat. At last, turning his attention to the 10 feet of hardwood lying along the gunwale, Jose asked, *"Listo, Patron?"* I glanced at the heavy harpoon, the hand-forged point glinting from its fresh honing. The silky, white nylon sashcord I happened to have brought along was secured to the ring welded below the steel barb. It ran along the shaft six feet up the green wood to a self-release knot and trailed off to a neat coil in the dugout's bow. The rig was ready. *"Listo,"* said I, indicating that I was ready too. Somehow, this didn't seem quite as good an idea as it had yesterday, or even an hour ago. Jose mumbled something in Spanish about his spiritual prospects, crossed himself and swung the bloody sack of cattle guts (bought from a rancher a couple of miles from camp) over the side and into the murky water.

A liquid highway of red poured downstream from the sack, snubbed up to the side of the *caycua*. Thirty seconds went by before we saw it; starting with a small riffle on the surface, rising slowly until easily two feet of the fin gleamed like wet lead above the long, tapered shape. It tacked across the water as the big shark cast its 10 feet back and forth like a pointer trying to pinpoint a crippled pheasant. Other dorsals appeared as the bag continued to seep its bloody invitation into the current. I slipped the leather glove over my left hand and raised the harpoon as the shark crossed the riffles directly for us, the long upper lobe of the tail following the dorsal.

Closer, it glided to the dugout, the half-open mouth agleam with snaggled teeth, its cat's eyes dully reflecting as it paused in a patch of sunshine a dozen feet away. Wait; wait, goddammit I kept repeating to myself over and over, but at this range, I could stand it no more. Jose's shout of *"Ya!"* melded with my thought of "Now!" and I threw with every ounce of strength I had, aiming for the head, hearing the *thug*! as steel met gristle and muscle. For a fraction of a second, nothing happened. Then he charged.

I was still off balance from the throw, my weight forward as he smashed into the

(24)

SHARKS

side of the tippy *cayuca*. The long harpoon shaft, jutting from the side of his head, flashed toward me and clouted me hard across a temple. I felt myself teetering over the wounded, blood-enraged shark, my balance going. Great Godamighty! I was going to fall in on top of him! Then, as if from nowhere, Jose's hand flashed out from behind me, snatching at my belt, jerking me back and down as I fell hard into the bottom of the canoe. Nothing ever felt so good as those skinned elbows and barked knees. I had nearly literally been snatched from the jaws of fate. We almost swamped the light craft, but somehow didn't go over. If you think that wounded ten-footer would not have bitten me, you might be right. But I suspect the odds are overwhelming that he would have grabbed me for sure.

By this time, the shark had turned and was tearing off downstream, the harpoon line melting from the coil of rope in the bow. I snatched at the nylon with my gloved hand to try to slow his escape, but it was like trying to snub off a falling safe. There was just no holding him. The balsa wood float tied to the end of the line whipped off the bow and skittered across the surface like a giant, terrified waterbug. Jose hoisted the anchor and we followed, catching up with the float a half-mile away, in quiet, deep water. I plucked it from the water and felt the throbbing power at the end of the line again, slowly, I worked him in only to have him take off again, the cord ripping through my glove so fast we had to wet it! There was a mark across the palm just as if a hot branding iron had stamped a straight line on it. At last, after almost an hour, I worked him close enough to let some rain through his roof with a burst from my semiauto 10/.22 Ruger carbine. *That* definitely took away some of his enthusiasm for continuing the competition.

Slipping the big gaff into the gill slits of the killer, we tried—for reasons I cannot imagine—to haul him into the *cayuca*. Fortunately he was too heavy—maybe 500 pounds—and I wasn't that hot on sharing the accommodations, anyway. It took almost two hours to drag him against the current back to camp, but I was anxious to have a look at the stomach contents. Already, after being gaffed, I could see that the shark was a female, lacking the "claspers" flanking the vent of the male.

It took most of our staff to drag her onto the shore, where Jose let out most of her sawdust with a butcher knife. Jose was a man who enjoyed his work. We took a look into the four-foot incision and got quite a surprise. There, each wrapped in its own little membrane bag were six very alive pups, thrashing and snapping. I unzipped one of the water filled bags and grabbed junior by the tail. (And I'm writing books about man-eaters!) As if he'd done it every morning of his life, he whipped around and buried a very efficient set of teeth in the edge of my arm. It hurt like hell as the blood flowed down my wrist, the teeth digging deeper with each jerk of the thrashing of the three pound body. All present, of course, thought this uproariously funny and it took a few very select phrases in my emergency Spanish to get somebody to

(25)

·stop rolling around on the ground and cut the infant's jaw hinges. After a bout with the all-too-familiar permanganate and lots of gauze, I hied myself off to check the therapeutic qualities of alcohol under field conditions. I am pleased to report that I was back on the river late the next morning, hardly noticing the pain of my genuine case of shark bite for the throbbing of my temples.

We killed six sharks the next day at the cost of both harpoons, terminal rope burns and about three miles of sashcord. I did, on the other hand, pick up some extremely handy phrases for my growing Spanish.

I thought at the time that it was strange all the sharks were female and near delivery. In most cases, birth would have occurred in a matter of hours if not days. Although only some remote reference at this date had been made as to what was then still *C. nicaraguensis* and not yet *leucas* birthing in either the lake or river, this deep area near the El Toro Rapids was clearly a nursery area such as has been recorded as customary with bull sharks in salt water. Maybe, in light of this dona- tion to science, they'll call it Capstick's Pool. For sure, they'll be naming no Sunday Schools after me.

A few days later, a well-known photographer came to camp and we were able to go through the spearing business again, with some pretty good still photo results. I regretted only that I was not able to find as big a shark then as the first one I stuck. In fact, when some of these action pictures of me were shown back in the U.S., they spawned a segment of ABC's *THE AMERICAN SPORTSMAN* back in 1966 or 1967. They starred Jack Nicklaus, the great golfer, spearing sharks, but it was one of those ventures where nothing seemed to go right. I haven't seen the segment nor have I heard of it being repeated. As a matter of fact, I came down with flu and was not even able to make the trip as a technical advisor, which was probably just as well for all involved!

As far as the two types of sharks supposed to be in the lake and river, one having a white the other a pink belly, I believe that I may have found the answer while spearing. Often, when feeling the steel, these sharks would roll over and over in the harpoon line, trying to presumably cut it. In the early stages of the fight, the belly would appear snow-white, which would seem to indicate a *visitante*, run-in from the Caribbean. Yet, when gaffed, it was noted that the exertion of the fight against the line had ruptured a layer of small blood vessels just under the belly skin, turning the stomach a noticeable, if not vivid, pink. Perhaps it's the answer, maybe not.

There is one aspect to the fresh water sharks and other fish of Nicaragua that does not seem destined to be solved quickly, though. If you recall, I mentioned that there are two lakes, Nicaragua and Managua. They are presumed caused during the same cataclysm, yet there is an odd difference between them: Lake Managua does not contain—and apparently never did contain—the salt water species. They are connected by a river called the Tipitapa, which, although not a reliable connection, without question links the lakes in times of high water tables. Yet, the sharks, tarpon, sawfish, etc., will not enter the slightly higher Lake Managua.

The bull shark probably kills more people in India, in the Ganges, under the *gangeticus* pseudonym than anywhere else. The number of bodies that become shark food alone must be staggering. Not just the Ganges, either: In 1959, 35 people were killed or mauled by sharks at the mouth of the Devi River, also in India.

Perhaps, considering the vast varieties and numbers of sharks to be found in salt water, you may consider this an overexcursion into what is really a sideshow of this great eater of men. For my defense, I say no. By discussing the possibility of death in so many fresh water lakes and rivers from what is so often thought of as strictly salt water risk, I have tried to simply hammer home again the point, the most important thing any human should realize about sharks: *count on absolutely nothing.*

CHAPTER TWO

LIONS

To get one thing straight between us right off, the object of this chapter never really was to make you read between the lions. That's just the way things worked out.

If there were an especially distinguishing characteristic peculiar to the African lion as a dues-paying, card-carrying member of the feline man-eating union, it would be his heartwarming sense of comradery; the same wonderful tendency toward social togetherness and singularity of purpose as found in a school of sharks at the height of a feeding frenzy. Animal behaviorists and other people who tell the general public much of what they really don't want to know about wildlife and its eating habits generally refer to lions as a "social" species, somewhat dulling the less palatable facts of life and death in the wilderness by the implementation of a term that rather seems to me better suited to a description of a gaggle of gabbling grannies pitching in to share the stitching of a quilt or the tossing of a few martinis over some duplicate bridge rather than acting as cooperative henchmen in the hunting and killing duties that have lead to the greatest aggregations of dead people at the hands of animals ever recorded.

Now, before my conclusions are hastily formed that might unfairly besmirch the reputations and accomplishments of many single and extremely talented individual man-eating lions, it must be observed that these loners probably constitute nearly

90 percent or more of the "injured" category of man-eaters discussed earlier. The reason for this logic is that it would stretch the laws of probability to the twanging point for more than a single, aged or otherwise infirm, lion to join forces for the common good with another disabled lion for any purpose other than eating each other. There have been, to safely cover the "fudge" factor of 10 percent I built into the last observation, some very deadly exceptions that, are "the exceptions that prove the rule," whatever in the world is meant by that charming old ambiguity.

Nonetheless, despite the nonconformist individual man-eating lions who so crassly make a name for themselves by acting alone and running up some quite impressive totals of kills, the factor that makes the lion, or, more especially, a particular and usually related group of lions so incredibly dangerous a threat to the human population of what may be several thousands of square miles is the simple fact that there may be from a pair to a dozen or more full-time activists. So, it's one thing to concentrate on the killing of a particular tiger who may have collected even a couple of hundred human beings or a leopard who has done the same. That's tough enough, to be sure. But, at least the hunter knows there's only one who, unless the laws of physics have been rewritten, can only be in one place at a time. Not so with old *Simba*. He'll not only outsmart you; he'll simply outnumber you.

This peculiarity, or, perhaps better said, individual characteristic of *Simba* as a man-eater is almost surely a direct reflection of his standard game hunting tactics which involve a great degree of cooperation between members of a socially or biologically related group of lions. This, incidentally, brings to mind a point which is often mentioned as being an infallible truth about lions when stalking and killing game. Most people, possibly because of a documentary film produced some years ago, are under the impression that the females do all the killing of game for the males. True, the male will often drive a potential victim toward a female in ambush, especially in the case of domestic stock, by going upwind of the prey, grunting, urinating and giving other previews of coming attractions sufficient to cause cattle to break out of their pens and rush smack into mama's attack plan. But, it has been my experience that nearly as many kills of both game and stock are by males as by females. Perhaps tactics vary from area to area, but you couldn't prove it from my observations.

It's interesting the way most things in life have a way of swinging pendulum-like from one extreme to another, be the matter hemlines, politics or the popularity and public awareness of man-eating lions. Even a couple of decades ago, the most famous lions of all time (before the hemorrhaging heart of Hollywood decreed otherwise) were not the cuddly Elsa and her Jack Paar supported doubles from *BORN FREE,* or even the pathetic cross-eyed Disney bloke, but a pair of human killers called the Man-eaters of Tsavo. The reason they were all the rage was not just that they up and ate anywhere between 28 and 100 coolies imported to labor on the

Mombasa-Victoria-Uganda Railroad at the turn of the century, but that they actually stopped the building of the line at the Tsavo bridge, 132 miles up the coast. Give the devil his due, I suppose, because the Victorian British Empire was no bunch of *wallahs* to fool around with, particularly when its wrath was unsheathed in the shining khaki form of Lt. Col. John Henry Patterson, D.S.O. Not only did Patterson probably convert thousands of the most definite atheists on the basis that there simply had to be a God for his blundering not to have gotten him eaten, but J. H. even wrote a hair-raiser about his death struggle with the savage brutes. It was an instant best seller (why don't lions like that ever come *my* way?) and conferred upon both the author and the lions eternal berths in the celestial Halls of Montezuma—at least until the "I-Raised-Lion-Cubs-and-Returned-Them-To-The-Wild" literary syndrome temporarily edged them from center stage.

Personally, I absolutely treasure the bad old days. And, from the amount of more recent magazine exposure to the earlier facts of the more realistic culinary arrangements naturally extant between *homo* and *leo,* it would appear that I'm not the only one who realizes that there is a great resurgence in interest in the days of African derring-do much like that which Time, Inc., is now promoting on the realities of the old American West. The only sorry thing about the African aspect is that the damage is probably irreparable through loss of habitat rather than game depletion through hunting as sentimentalists prefer to believe.

That many of the outstanding features of the old Africa are still very much with us is well borne out by the killing and eating of my ex-boss, Peter Hankin, by a lioness in the Luangwa Valley of Zambia in 1974. I have told of Peter's classic shucking off this mortal coil in my first one-man Comanche raid on literary decency, *DEATH IN THE LONG GRASS* back in 1978 and won't repeat it here but to say that Peter, a famous professional hunter, was killed with a single bite in the neck by a famished lioness on what would have been Labor Day in America that year. While his clients on photographic safari in nearby tents listened, he was partially eaten before the lion was wounded by rescuers the next morning and killed a short time later by my friend and hitherto colleague with the old Luangwa Safaris, Joe Joubert. I have also told of the contemporary man-eating incidents involving Len and Jean Harvey, Willie DeBeer, Colin Matthews and a score or more of African blacks, none of which I have ever seen a wire service report upon, save Joy Adamson. Speaking of whom. . . .

Now, *there* was a mighty odd incident that I think bears the sharing of a few thoughts. It has always struck me as peculiar, as in the deaths of many of the people above by man-eaters of any type, that very little if any news of such shenanigans ever reaches the outside world. The only report I ever saw on the rather spectacular and certainly newsworthy death of Peter Hankin appeared two years after the event

(31)

in Edward R. Ricciuti's book *KILLER ANIMALS*, whose author happened to actually be in Zambia at the time of Peter's killing. Hankin's name is not mentioned nor are any details beyond the most obvious; i.e., he was killed and eaten while sleeping in his tent. If there were any other reports, perhaps I simply missed them. Still, it seems unlikely that they would not surface with more obviousness. Or, are they deliberately suppressed in countries where wildlife-orientated tourism is so important? Methinks there may be something to this, or perhaps it's just the lack of reporters hanging around remote African villages waiting to file copy on the next person ingested by a hungry lion. Still, the case of the death of Joy Adamson, author, animal lover, behaviorist and sentimentalist is especially odd, at least from the information available in the press.

Before making any further comments on the physical matter of Mrs. Adamson's demise, I wish to make clear that my objection to her views and work is not personal but a matter of emotional interpretation; by which I mean that I have no question of her sincerity in her ideas expressed toward the preservation and protection of wildlife. As I understand it, all proceeds from her books were donated to "The Elsa Wild Animal Appeal" which, however effective with a title like that, must have meant well. My argument with the attitude of people such as the late Mrs. Adamson lies in the unrealistic sentimentalism which has lead tens of thousands of otherwise unexposed people to believe that lions, if not all the great carnivora are, in reality, poor tabbies starved for the godlike gift of human affection. Would it were so, yet I cannot but believe that, indirectly of course, every kid who has his face torn off by an irate black bear in some American park or each damnfool who has seen *BORN FREE* and is mauled by a lion who doesn't purr on cue or appreciate the human concept of affection is at least to some degree the fault of Mrs. Adamson or Mr. Disney.

I first heard of the death of Joy Adamson from the wire services through the *Miami Herald* of January 5, 1980, which, in part, reported the following:

AUTHOR OF 'BORN FREE' KILLED BY KENYA LION

From Herald Wire Services

NAIROBI, Kenya—Joy Adamson, whose *Born Free* awakened millions to her beliefs that "once wildlife is gone, it is gone forever," was mauled to death by a lion, it was reported Friday.

Friends said that Mrs. Adamson, 69, had taken her customary evening stroll Thursday night in the bush outside her tent camp in a remote area of northeast Kenya, and came across a lion chasing a buffalo.

The lion turned and killed her.

She was found by a park employee, lying face down with heavy wounds on her hands, arms and head.

"The lion is still at large, but we have tracker teams out looking for it," a police spokesman said.

Her body was found about 1,000 yards from the barbed-wire enclosed compound where she had been conducting experiments in returning tame leopards to the wild.

The leopard experiment was similar to the lion project that touched hearts around the world.

In 1956 she adopted Elsa, the wild-born heroine of *Born Free* whose mother had been killed in self-defense by Mrs. Adamson's husband.

Believing, as George Adamson once said, that "the mentality that condones wild animals in lifelong captivity is little removed from the mentality that condoned the slave trade," the two conservationists raised the cub on love and affection while also training her to hunt.

Now, that seemed pretty pat to me. No question at any time of any kind of human foul play, although I commented to several people at the time that it was going to be *most* embarrassing to the "Elsa Wild Animal Appeal" for the world's most famous lion lover to have up and gotten The Big Chop from one of her own sweeties. And, what of the dozens of other heavy fund-raising organizations who generated millions of cool, green dollars for the sake of all those lovely, soft-eyed endangered cutie-pies? What would good old Cleveland Amory have to say about such a turn of events? Clearly, the death of Joy Adamson at the hands—excuse me, paws—of a lion would not help anybody's fund-raising let alone their projection of the African lion as lying characteristically with lambs in a bed of jonquils.

My, my! Guess what happened? Seems that all those tracker teams spooring up the lion that killed Mrs. Adamson must have been mistaken. Apparently, the mention of a buffalo was all wrong, too! From an apparently confirmed story of death by lion, the "official" verdict was now that Mrs. Adamson had been murdered by a human since some of her belongings were missing.

You may have noticed that you never heard much more—if anything—about the matter. That, in itself, is as weird to me as the abrupt change of cause of death. I wasn't there and accuse nobody of knuckling under to any big money preservationist sentiment-venders simply because it would have raised a few donor eyebrows for the author of *BORN FREE, FOREVER FREE, LIVING FREE* and the rest of the stuff to have been killed by a lion. Heavens no! I guess the murder of a famous authoress in the midst of the African bush just wasn't interesting enough for the press to pursue. Don't you think?

As a man-killer, presuming the victim has the giddy good fortune to be killed by the lion before starting to be eaten, which is not always the case, *Simba* is only at his best when he has a chance to size matters up properly. The best time to get yourself killed by a man-eating lion with the least complications is while sleeping. A set of thumb-thick fangs will clamp over your temples or whichever portion of your skull is handiest and you don't give a damn from then on whether the taxes get paid or not. It's all these folks that either toss around just at the critical instant or wake up and try to struggle that give even accomplished man-eaters a bad name for clumsiness. Bitten through the shoulder because of a thoughtless lurch at the last moment normally means being dragged off to an extraordinarily unpleasant several moments. Harry Wolhuter, who in 1902 managed to stab to death a man-eater who had pulled him from his horse and dragged him 60 yards before Wolhuter performed some impromptu surgery, did not recommend the experience unless absolutely necessary.

The tiniest percentage of men who have survived lion wounds received them from a man-eater. Collections of keloids accumulated while hunting ordinary lions, for example, don't count for this book. Of all the victims of the Tsavo lions, I recall only one of those attacked who was actually wounded and survived, that being District Commissioner Whitehead, whose native sergeant, Abdulla, died in his place mere feet away. Whitehead sustained a moderately clawed back but recovered. I have twice been knocked down by man-eaters, but, through what I can call only sheer peevishness, they left me not a single scar worth a belt of even a tertiary brand of scotch at a safari lodge bar. The bruises, to be completely fair, were really beauts, though. And, then there was John Welford. But, I think I'll save him for later. . . . Lord knows, he did as much for me!

Seems that we keep getting back to two areas of peculiarity about man-eating lions; first, that they work in well organized relationships with each other and that, second, there are a hell of a lot more of them wandering around the present and relatively recent past who, for reasons that may seem a touch creepy from our points of view, are quite real to the majority of people affected. I can think of no better example of the man-eating lion at his most destructive and disruptive worst (or best, if you're on the home team) than the more than 15 years of absolute terror generated and damn near perpetuated, as the Disney writers might have put it, by a happy-go-lucky pride of fun-loving felines guaranteed to make your sides ache with their antics. Undoubtedly, they caused even more aches than that. These were the man-eating lions of the Njombe district of Tanganyika (now Tanzania) and although you probably never heard of them, they killed better than two thousand people! Yeah, two thousand. Now, you have recently noticed how upset a heavily populated country like the U.S. gets over the possibility of losing just 53 hostages. As I write this, the Atlanta child-killer or killers has just cut his 23rd notch. Hell, Berkowitz, the "Son of Sam" killer barely got into double figures and we still hear plenty about him. So, you ask—with good reason—how can a lousy bunch of lions kill and eat the equivalent of four Boeing 747 plane loads of human beings and you've never even heard of Njombe?

Well, you're probably not listening. And, you're not listening because nobody's talking. Call them werelions, say it's black magic. Lay down your thin and dusty dime and take your choice. Here's the story; it isn't pretty but it sure is weird. . . .

Considering the awesome numbers of dead people as well as the years over which the events took place—not to mention the well, uh, less routine aspects of the whole matter, it's difficult to decide where to begin the saga of the Njombe lions. But, since the animals themselves, if that indeed is what they were, are the one factor common to the whole episode, we might as well start with them.

We don't know precisely how many man-eaters there were at any given time in the Njombe district except to note that 22 lions were killed over the last year or so of their activities, any subset of which would include the total of man-eaters along with what must have been a few "innocent bystanders." Both hearsay and records of the Tanzanian civil service indicate that the first wave of people-eatings began about 1932 and under circumstances apparently less than worthy of inclusion in the public record. To paraphrase District Commisioner W. Wenban-Smith, who took over the government post at Njombe, at the southern edge of the man-eaters' territory after World War II, some lioness unable to fend for herself must have taken up snacking on the local gentry and passed the practice on to her cubs, who themselves spread the idea that black is not only beautiful, but delicious. Within a short time, say a year or so, in the roughly circular area contained within the village of Njombe to the south, the Great Ruaha and Madowi Rivers to the north, the Poroto and Kipengere mountain ranges to the west and a seemingly arbitrary north/south line forming the eastern border of activity running just east of the hamlet of Matipu and the village of subchief Jumbe Musa, about one person began to meet death every three days on the average. This doubled in the late '30s and during the war years; at least 600 people having been killed in three villages alone between 1941 and 1945! On the off-chance that you don't happen to have a road map of southwestern Tanganyika in your glove compartment, just look slightly northeast of the corner where Zambia, Malawi and Tanzania come together in your Atlas. At the edge of the mountainous highlands is what's called the Buhoro Flats, and it was in this great sun-seared snarl of thorn-fanged *nyika* that most of the festivities were conducted. Straight through this whole vegetable obstacle course passes the brilliantly named "Great North Road," which runs east/west through this portion of the country and in only two seasons provided 17 street sweepers as lion food.

Probably the most enticing coincidence that was to run through the bloodstained fabric of fact and fancy is the birth of a legend which oddly began to spread almost exactly at the same time as the lion killings themselves began.

Perhaps a word is in order concerning African superstitions and legends. First off, if a superstition is strong enough to get started in Africa, then it's probably tough enough to give steel-belted radial mileage. Secondly, the bouillabaisse of black magic, lycanthropy, cannibalism, ritual murder and mutilation (just for starters) is rather different from quaint western notions of leprachauns and faerie queens; walking under ladders and touching wood. I hastily suppose it's that life in the bush is by definition a lot closer to the realities of death than we experience in our AC-DC microwave culture. Rarely do we wonder how loudly the tiny lamb gargled his death-bleat during the first step in supplying a succulent rack of chops. Our eyes do not grow misty with much more than the financial shock of parting with the where-withal to purchase a pair of fine calfskin shoes. I do not find myself contemplating an omelette as just another brace of unborn chickens who will never go to college.

Not so the African; and I do not use the term racially. It is and was so with every close-subsistance culture. Hell, when I was in central Africa in the 1960s—that's this century, not last—as a professional hunter, the average man working for me had some ten or twelve children. It was tough to face, but if any more than two of them had survived under normal circumstances by avoiding infections, falling into campfires as infants, assorted biharziasis, sleeping sickness, snakes, poisonings, fights and the rest of the risks of life, then the whole lot would have starved to death. Not a pleasant prospect, especially if you have a conscience. Let's just say that the facts of survival vary from place to place. The New Yorker tries not to be run down by a taxicab. He doesn't have to worry whether the cab will eat him if it can catch him. So, the African, helped along by the brighter of his brethren who recognize profit in fear and the manipulation of others, sees goblins and ghouls behind every bush and tree, a werelion or werehyena in every shadow of the moon in the dense bush and witchcraft so thickly all about him, to this day he carefully hides his ex-creta and fingernail parings in remote areas that they may not be used against him in a curse. In some places the highest form of respect one man can show another is to spit into the palm before shaking hands, the point being that the spitter trusts the man with whom he shakes hands so much he offers his own spit in hopes that such a deadly substance will not be used against him. To many rural Africans such a ges-

(36)

ture is rather like handing an armed enemy your only spear just to show that you trust him as a peachy keen guy.

It was against his background of ingrained terror and apprehension that the subtle legend of Matamula Mangera, headman of a village called Iyayi on the Great North Road began to seep through the bush as the killings of the lions spread across the Buhoro Flats. A well-known and feared witch doctor, Matamula had been deposed by a superior chief under accusation of one or another form of corruption a bit more involved than practicing black magic without a license. Even the most casual study of his personality and tactics would convince anyone that Matamula would have done very well in the Roman Senate under Caligula and might have even survived politically in modern Washington, D.C. He never made a direct threat, but saw to it that word was passed: if he was not reinstated, more and more of the people would be taken by the *"dudu ya porini."* This means, literally, "insects of the bush" in KiSwahili, but would translate better into English as "bush critters" or "varmints." As the wave of man-eating went on, it was further believed that Matamula had developed himself a proper man-eating menagerie and staff of attendants. One henchman by name of Mkakiwa was the lions' herdsman and lived with them in some secret place between the villages of Igawa and Rujewa, and another stooge named Hamisi Sayidi was supposed to actually handle the lions and point out their next kills to them.

Part of the reason for the absolute belief in such arrangements by the primitive local tribesmen (not to mention a healthy respect espoused by the local missionary in charge of the Lutheran station, Reverend Martin Nordfeldt, who suspected werelions) was the supposed recent outbreak of lycanthropy or the magical changing of people into animals and back again at will, to commit murder. A plague of lion-men, *watu-simba*, actually a collection of killers forced to insanity through drugs and long confinement and hired out by witch doctors for vengence and other murders under the guise of lions, had recently been quelled in the nearby Singida district to the north and the logical carry-over in the minds of the Njombe locals was that this was merely another helping of the same.

So, the situation we have accumulated after fourteen years of constant man-eating between 1932 and 1946, when the game department of Tanganyika finally began to concentrate on the problem through the pleading of Wenban-Smith who telegrammed that conditions were "pathetic" is more or less the following: 1500 square miles of relatively high population density Africa is so completely terrified that should a parent lose a child during the night to one of the man-eaters, neither mother nor father would dream of mentioning the fact to even a close friend for fear that the simple invocation of the man-eaters by name would suffice to place the speaker next on the list. As far as the outside world was concerned, there were no man-eating lions of Njombe. They were simply never spoken of. And, these are precisely the problems that George Rushby hit head-on when he took over the toughest job of animal control in Africa in 1946.

If you happened to be a marauding, crop-raiding buffalo or elephant, a re-calcitrant rhino or a man-eating lion or leopard, the one thing you could do nicely without would be George Rushby, game ranger of the Tanganyika game depart-ment. Born in Nottingham, England in the second month of this century, Rushby was raised by a tough stepfather whose cronies, if nothing more constructive, taught him rough-and-tumble fighting good enough to scratch out a carnival living when he pitched up penniless in South Africa in 1921. Between those years and the time he came up against his greatest challenge with the Njombe lions, Rushby had been a professional trader, poacher, ivory hunter, gold digger, bartender, bouncer, planter, safari *bwana,* guano prospector, and, finally a ranger for the game department. He was apparently pretty good at all of these, too.

After the second war, the end of which was celebrated to a sufficiently soused conclusion to force the sodden ranger who preceeded Rushby in the Njombe area to retire or die of the "rats," George and his small family took over the Southern Highlands Province under the command of Victoria Cross winner Monty Moore. It was Moore who advised Rushby of the "odd" aspects of the Njombe lion killings, which immediately got the attention of the transferred ranger.

The first thing Rushby discovered upon reconnoitering the Buhoro Flats on ei-ther side of the Great North Road was that, except for the White Road supervisor, all of whose workers had been eaten, nobody else would even mention the existence of the man-eaters, so terrified of retribution was the entire surviving populace. To make matters worse, and on the solid logic that if you believe in God you must believe in the devil, the Reverend Nordfeldt was also convinced that although they might look like lions and leave lion pug-marks, that was as far as the similarity went. Only a couple of nights before Rushby arrived at the mission, one of the station's black employees was just rounding the edge of the church at dusk with a lantern when he disappeared in a shriek of terror. Nordfeldt and the rest of the brethren came hustling with lights and plenty of artillery only to find the disem-bowelled—and uneaten—body of the man a few yards off the path in a cornfield. The Reverend had gotten the idea that this was in retribution for one of the ex-tremely rare survivors of the Njombe lions' attacks a week before this. A man had been grabbed in bright daylight, but in carrying him off, the lion somehow jammed a sharp stick or thorn into his own face and dropped his dinner. He was the only survivor anybody knew of, and this was presumed to irritate the lions.

Incredible though it might seem, at least nine years passed without any attempt whatsoever to kill the man-eaters from their first activities in 1932. Dusty Arundell, the ranger in charge before Rushby, seemed always to be called off on other, more important official business whenever he was about to open warfare with the lions and his large staff of armed and well trained native department personnel were themselves so frightened to even breathe the word *Simba* that the idea of actually trying to kill one of the supernatural lions was completely unthinkable.

In the first years of the war, a mission in the lions' area called Kidugala had been turned into a Polish refugee camp overseen by a British commandant named Wagner and staffed by Italian prisoners of war who happily performed the maintenance work. As Wagner started to realize the astounding numbers of local people being eaten all around his mission, he began to sit up at least two nights a week in a tree platform in hopes of getting a shot. With him usually lurked two of the Italian prisoners who had been keen sportsmen before the war, men who were obviously closely trusted as they were allowed firearms. Although they spent many weeks waiting for the lions, it's bleakly interesting that the only time the killer showed up was one night when Wagner was away from the mission on other business and the Italians had decided to give it a go by themselves.

It was 1 a.m., the night chill and quiet under the slice of moon drifting soundlessly between patches of whispy cloud like a flashlight beam filtered through surgical cotton. Lulled by the peacefulness and warm comfort of their blankets and thermos flasks, the Italians were practically petrified with shock when a ferocious grunt followed by a snarl like a chain saw tore a ragged hole through the night directly below them. With a hard shake, the tree swayed as one of the lions tried to climb up into the platform to reach the men. That was all, brother! Down came a hail of dropped rifles, flasks, flashlights and blankets onto the lions as the Italians took off from the platform higher into the branches of the tree, bug-eyed with terror. In the morning they virtually had to be pried from their death-like grips on the smaller high branches when a team of rescuers arrived. The first attempt to kill the man-eaters rather set the pace for the next few years. Neither Wagner nor the Italians is recorded as having tried again.

The only—or at the time, the first—African to dare to fight back against the Njombe lions was an inhabitant of Rujewa, a village just about in the north center of the lions' range on the Buhoro Flats. It was early of an evening, and the man's wife was one of the several women in the "street" between the huts of Rujewa. In a blur of savage motion a tawny shape burst from the bush and knocked over several of the women, sinking its fangs into the man's wife and dragging her easily off, probably already dead. Half insane with rage and horror, the man raced for his hut and grabbed his Tower musket, a blackpowder affair that is quite reliable, I can assure you, having shot one a great deal in my youth. (No, the Tower musket and I are not contemporary!)

Screaming for his friends and neighbors to help him the man realized that this was not volunteer night and, so, went on his own bravely toward the sound in a nearby thicket where an unknown number of lions were dismembering his wife into more easily handled chunks. As he closed on the place, a lioness stepped out of the bush and began to walk slowly straight at him, exactly the way a menacing human attacker would, stiff-legged and calculating. The last gleaming of light shone on the sheet of gore covering the lioness' chest and face, and, in her jaws was clenched one of the wife's legs, severed as if by an axe stroke. At a few yards she stopped and stared back over the musket sights at the man, her eyes icy in their challenge, daring him to fire. Numb with a horror that seemed to wash over him in waves, the man began to tremble, shaking until the quivering musket rattled so that the percussion cap fell off the nipple and the gun was useless. With a grunt of disgust, the man-eater slowly turned her back and walked away. Nobody tried to shoot any holes in the Njombe man-eaters for a long time after that.

As the years wore on and whole villages were decimated, there was possibly only one man of any importance in the whole district who was not completely in awe of the supernatural properties of the lions and their rumored master, Matamula Mangera. The man, named Jifiki, was the young subchief of the village of Wangingombe on the Great North Road, a small place with an intriguing name meaning "The Place of Stolen Cattle." If not immune from all the so-called mumbo-jumbo, he was at least much less affected by it than most others and provided Rushby with much of the historical background of the outbreak from the African point of view. Probably the most accurate description of Jifiki's view was that he didn't really believe in the black magic of Matamula and his lions, but he wasn't about to take any chances either. In other words, there's no such thing as black magic unless, of course, there happens to be.

Rushby knew better than to go to see Matamula personally as his doing so would be interpreted as acknowledgement that the man's power was genuine and recognized by the "*Bwana* Game," as all rangers are called. Instead, Rushby kicked the collective butts of his rangers back into the Twentieth Century and started drawing plans for an extended campaign against the lions that would conclusively prove that the only thing supernatural about these Njombe cats was that they had lived so long already.

No sooner had George gotten organized than, as if by eerie coincidence, he was ordered to drop operations against the lions immediately and proceed to the distant Rukwa Valley to work on an outbreak of locust with an international organization. As he loaded his vehicle and drove off, it struck him that this was the same thing that had happened to poor old Dusty Arundell each time he started to wage war against the man-eaters. Rushby began to get an odd feeling in the back of his neck.

George was away for two months doing whatever one did in 1946 to slow down an infestation of locust, not able to return to hunt the man-eaters until January of 1946. Picking six of his least superstitious and most reliable African game scouts, he broke this crew into three groups; two men at Jifiki's Wangingombe village, two on the northeastern border of the man-eaters' territory on the Great North Road at a village run by headman Jumbe Musa, another non-spooky type, and the remaining two free to chase rumors of the lions' presence and run a string of set-gun traps. As is almost supernaturally typical with man-eaters, though, three weeks went by with the standard compliment of people eaten but not the slightest sight of the lions was seen nor was one trap touched.

As the weeks stretched into nearly a month, Rushby began to understand that either the Njombe lions were extremely lucky or they were smarter than dammit. No sooner would they gang up and make a kill by tearing through the roof of a hut or two than they would eat the body and scatter to the four winds, acting almost as if they had been surprised on returning to a carcass early in their man-eating career, which George believed to be untrue. Whatever, these tactics sure put a hell of a kink in any attempt to apprehend the cats as they never hit the same place twice in succession nor ever remained in the same part of their range after a kill. Meanwhile, word continued to spread, reinforced with every eaten corpse, that the terror would never stop until Matamula was returned to power. Exhausted and frustrated, Rushby and his men decided to retrench and consider their tactics.

The Great North Road and two smaller tracks to and from Njombe were, Rushby realized, the best arteries for the accumulation of information on the lions' whereabouts and grisly activities. So, he set up an intelligence gathering camp on the main road and tried to second-guess the man-eaters from there as word arrived of new killings. With its usual sense of ill-timing, the game department forced George to return to his office for a week's quintuplicate paperwork before he could implement this new approach.

After a week to think over his first efforts, he was even more frustrated than when he left as he realized that his failure to even get a shot was interpreted by the local people as only more proof positive that the lions were supernatural. As news poured in each day, Rushby found only chewed-over scraps of what had been people after the lions had finished feeding. A trademark of these particular man-eaters was their love of human brains, every victim's skull carefully bitten open and licked clean of its contents. As is also very often classic with confirmed man-eaters, they would eat nothing *but* humans, disdaining cattle, goats, even chickens in their fondness for dark meat. Even the village herds of cattle seemed to lose their fear of lion scent as if knowing that it would be the herd boys who would be killed and eaten. One odd quirk, Rushby noted, was that the lions seemed to have some kind of grudge against wild pigs. Time and again, their bodies would be found on the lions' trail, stomachs ripped out and uneaten; killed it would seem, for some grisly sport.

The strain of night hunting the man-eaters from, first, a freezing perch in a tree and later from pits dug in the ground near the pathetic scraps of a recent victim began to tell not only on Rushby who, over three weeks blew one leopard, one pig, one jackal, two hyenas and somebody's stray goat to the incorporated limits of Perdition with his 9.3 x 74R mm double rifle (about the continental power equivalent of the British .375 Holland and Holland Magnum) rather than turn on his headlamp to verify a shadowy presence in the night and perhaps scare away one of the lions. Two of his best African game scouts of the elite six had to be removed from maneater duty after confidentially suggesting to Rushby that it would make things a hell of a lot easier to open negotiations with Matamula. The strain was building. . . .

Clearly, Rushby's intelligence system wasn't doing much good as the *dudu ya porini* had the most disconcerting habit of leaving the killing ground before George or his men could get there. The dry, rocky terrain made tracking almost impossible as the ground was as hard as an asphalt parking lot. Disgusted, he took a couple of days off to see if he could remember his wife's first name back in Mbeya, where his family was quartered, and figure out his next tactic.

Realizing that all but two of his scouts, Alfani (who was part Somali from the far north and once George's cook) and Fungamali, were scared green of Matamula and "his" lions, Rushby decided to work with just these two men in a concentrated "flying squad" and two more stout hearts who patrolled on their own following up news. After three days of practicing small group tactics with the two teams without any word of lions, a report came in with a transport driver of a multiple killing at Mambego, nearly fifty miles away. When the rangers got there, the place was still in shock, pieces of people scattered around a goat pen and the only live population were some old folks working hard at getting stoned on corn beer and a somber eight-year-old girl who had survived the night's attack by lying in frozen terror beneath some skins that had fallen over her when the lions tore through the roof and carried off one or both her parents. She wasn't sure if the remains in the goat pen belonged to either or both.

The little girl seemed to take to George, whom she fed from the abundance of abandoned food left in the village while he waited up for the lions to return, but knowing they would not be back. In the morning, the girl's father came walking back into the village. Her mother was, indeed, one of those eaten in the *kraal*. The girl's father was not anxious to guide Rushby and his scouts, but I suppose a good view of the open muzzles of a double express rifle may have jogged his sense of civil responsibility.

From dawn until noon the little group eased through the savage's hooked wait-a-bit thorn and bush until, with complete astonishment, Rushby looked up at the hiss of breath from one of his men and saw four lions standing in the shade of a tree at the edge of a thorn patch, watching the men with what must have been equal surprise. Without a pause, Rushby knelt and fired the right barrel into the nearest, a female, smashing her upper front leg. Lions exploded in all directions at the shot, the lioness hit spinning and biting at her wound. Rushby's second bullet pulverized her head, dropping her into a heap. His teeth gritted, he reloaded both barrels of the double and with complete concentration shot her twice again, ignoring the others. That was one lion he wanted extra dead.

The three Africans were pop-eyed with terror and astonishment, clearly expecting the man-eater to get back up, walk over and chew the *bwana* into handy sized portions. Rushby walked over after dropping another pair of the long cartridges into the chambers and inspected the lioness, telling the Africans to have a look for themselves. With definite reluctance, they did, seeing only an overweight young lioness with a glossy coat and in perfect condition. If she were a werelion about to change back into human form, she'd clearly better hurry up as the blood around her four bullet holes was already starting to dry. But, of course, the answer was obvious: she was a genuine lion, not one of Matamula's, shot by the *bwana* by mistake.

Rushby didn't give a damn. The point was that he'd actually killed a lion, man-eater or not, although certainly from her condition, she was one of the deadly pride. From this point on, the people of the district would realize that the real article was in the vicinity and could be killed without some weird curse to worry about. If this was so, it would encourage resistance instead of the terrible apathy that had for a decade and a half cost more than 2,000 people their lives, not to mention their bodies.

It still took considerable coaxing before even Alfani would actually touch the carcass, but when Rushby decapitated it, the Africans at last agreed to help with the skinning. When the body didn't turn into one of Matamula's "heavies" under the knife, they became more enthusiastic and even carried the skull and skin back to Mambego. Here, the people seemed to take a bit of heart and the surviving little girl presented Rushby with a dozen cooked eggs, secure in the belief that it was upon hen's eggs alone that white people lived. On his arrival at Mambego, Rushby got word from his other two scouts, who were patrolling near Jumbe Musa's village on the Great North Road, that the lions were eating people as if there were no tomorrow and to come at once. On his way there, George dropped off the head and skin of the lioness at Wangingombe and saw that Jifiki and his people were really heartened at the kill. Driving on to Jumbe Musa's, Rushby was further relieved to find that the extremely tense wording of the summons from his scouts was due to excitement, not fear, and that they seemed anxious to get on with the hunt rather than having lost their nerves as he had feared. When told of the death of the lioness, they were delighted and hot to hit the trail in the morning, filling George in on the details that night.

The scouts had been having problems with the local people, who were actually antagonistic to them for fear their hunting the man-eaters, however unsuccessfully, would bring down the wrath of Matamula. If they could only have a kill in Musa's area as George had done in Jifiki's, it might be enough to turn things around. Equally important, for whatever reason, the large number of lions eating the tax-payers at Jumbe Musa's were getting lazy and not moving so far or often after making a kill. The possibilities of the lions actually beginning to do something—anything—predictable was exciting enough to keep George and his men awake at night; but three days of dead blanks, weary, foot-sore and thorn-torn, soon had their spirits well deflated. And then, on the cool, bright fourth morning, it started to change.

The spoor was wide and clear, the three men trying to decide just how many lions had created the trail when they looked up at a snarl and saw a hefty, young male turning to disappear into the cover. The three rifles crashed together and the cat fell in a deflated heap. Overjoyed, the lion was skinned by the men and shown off at the village to the populace, who was suitably impressed. Rushby even went to pains to claim that it had been one of the scouts' bullets that killed the animal rather than some potential special magic of the white man's rifle. That night, smoking by his fire, Rushby made a decision: it was public relations time. In the morning, he would drop by for a visit with Matamula.

Over the perspective of years, it's interesting to wonder whether perhaps the Tanganyika game department was working for the lions and against Rushby. Whatever the case, before he could capitalize on his good fortune of killing two of the lions locally suspected as being controlled by Matamula, another signal was received which forced George to leave for the Northern Rhodesian border to replace a man dying of cancer and then down to the Rukwa Valley to bash some more locust. By the time he returned to the Njombe district, it was early April and much of the emotional weight of his planned confrontation with Matamula had evaporated. During his two and a half month absence, however, another lion had been killed by one of his scouts who was following a pride of the man-eaters although no discernible let up in the number of human victims was evident.

The fact that an African scout had killed a lion without any visitations from things that go *grrr* in the night had, in spite of Rushby's hopes, done little to alleviate the despondency in the area generally. He had best put off his confrontation with Mr. Matamula Mangera, still witch doctor-cum-headman, without portfolio. In fact, Rushby's return to try to hunt the werelions was not very well thought of by the locals who had recently sent a deputation to the Paramount Chief in a desperate effort to get Matamula reinstated so he would call off his "pets." Some white man mucking around with the whole arrangement was clearly not going to help the negotiations much.

Rushby spent the next couple of days hunting around Jumbe Musa's place and then headed down to Wangingombe, where he had similar negative results. Deciding to drop in on Wenban-Smith for a pow-wow at his headquarters at Njombe, Rushby was driving along, stopping at each group of huts or village he came to in hope of fresh lion news when he noticed one group of Africans, some 500 yards off the road, that acted strangely. There was a small cluster of spearmen and two men with muskets while the women had chased all the children inside of the huts. Asking what was going on, George was told that one of the men had seen a lion in a cultivation depression about an hour ago, but had lost sight of it as he ran to give the alarm. Not expecting much, George took his rifle and, with a couple of the spearmen, started tracking. Sure enough, there were pug-marks of a steadily traveling male lion which George and two of the spearmen followed while the musketeers stayed behind to guard the women if George's medicine wasn't as strong as Matamula's.

It was two hours before the brownish-tan dun movement flagged through the thorn brush as Rushby spotted the lion moving directly away at fifty yards range. Not wanting to only wound it and perhaps lose the cat, Rushby waited, following quickly, for a better angle for a shot. After five minutes, one of the Africans fell flat on his face over a gravel bank and the lion spun around with a nasty snarl, whipping into a full charge without the least hesitation. So fast was the cat and surprised was Rushby that the ranger failed to hold low enough a lead on the streaking lion and the first bullet passed harmlessly over the man-eater's back. On tore the lion, his mane bristling as if his tail had been plugged into an electrical socket, a hurtling mass of fangs and slashing claws that collapsed at only ten yards as Rushby's last bullet punched through his skull at eye level, the 9.3 mm soft point scrambling the cat's brains until they protruded from his ears. In a graceful, limp somersault, he looped through the air to land, dead as good intentions, on the ground. Rushby made another payment on the life insurance premium with a third bullet through the skull and noted that the body was similar to the others; shorter than usual but heavily built and with an extremely fine pelt.

Despite the fact that George stayed around the small group of huts hunting for three days, he was unable to really convince the people that this was one of Matamula's lions. In disgust, he continued on to Njombe for his meeting with Wenban-Smith. After their conference, both agreed that under no circumstances could Matamula be reinstated as this would be tacit recognition of the power of his black

(45)

magic and his lions, setting the district back a hundred years (as if it already hadn't been). There could be no more effective answer than the complete eradication of the whole pride of man-eaters, no matter how long it took or how difficult it proved.

As Rushby returned to hunting the Mambego area, he was intercepted by a note from his wife advising him of an important communication from the game department which he should return home to Mbeya to receive. Cursing the continued distractions to his hunt, he said goodbye to the little girl who kept feeding him eggs and headed back toward Mbeya. On the way, he determined to confront Matamula once and for all.

George pulled up at the village of Iyayi and called first upon the subchief who had been appointed to replace Matamula, a very respectable man named Ulaya. That Ulaya had been doing a crackerjack job of running his area was all the more reason that Matamula could not be reinstated. After offering his official respects, George wandered about the village speaking to the dispirited people until finally coming to the hut he knew to be Matamula's.

"Hodi?" called George in the standard address to a closed house.

"Karibu," answered Matamula, stepping outside. "Come near."

Continuing the conversation in the traditional KiSwahili sequence of asking after the crops, the cattle, the children, the weather and the rest of the drill did not shed a great deal of insight upon the character of the witch doctor, but it would have been an unthinkable breach of etiquette to say what was on one's mind directly. At last, Rushby was able to casually comment: "The insects of the bush are not having a very good time. Four of them are dead."

Matamula did not react but for a tiny smirk. With a casual shrug he answered with disinterest, *"Labda."* "Perhaps."

George drove on to his home at Mbeya where the important message awaited him. To the delight of his wife, Eleanor, the Rushbys were ordered on overseas leave for six months to England. Well, thought George, at least he'd be able to see two of his children who had been away for a full eight years, what with the war and schooling. That he would have to put off the pursuit of the lions was a bit disappointing as was the fact that he would have to sell his beloved 9.3 mm double rifle for spending money while on leave. His replacement would be the gangling ranger "Twiga" Rogers, named for the giraffe he vaguely resembled. George briefed him as best he could, suggesting he might try a mass beat or drive with a large number of natives as a possible way to kill a few more of the man-eaters. It was October of 1946 before the Rushbys saw East Africa again.

Twiga was waiting for him and Eleanor at Mbeya upon their return with depressing news. One additional lion had been killed by three Bena tribesmen armed with muskets, bringing the total of dead lions to five, but there had been no drop whatsoever in the overall number of killings. Further, the "drive" Rogers had organized had turned out to be worse than a fiasco and was in reality a resounding victory for the lions.

After a great deal of pressure being brought to a large assemblage of reluctant tribesmen, a "beat" was organized for three of the man-eaters seen near the village of Mdandu, just north of Njombe itself. With game department vehicles, the swarm of hunters armed with a motley array of artillery running from muskets to wired-up shotguns was transported into position and actually managed to find and surround the three lions in a bit of cover. Twiga Rogers was himself covering a point behind a low bank, which was just as well as things turned out. When the horde of gunmen opened fire on the lions' hiding place, things got a bit out of hand, the final tally being two Africans shot dead on the spot and a half-dozen more suffering from gunshot wounds of varying severity, including one which killed the recipient later in the hospital. The lions? They got away without a scratch, as, thanks to the bullet-absorbing bank, did Twiga Rogers.

In one small portion of the 1500 square mile range of man-eating activity, George was encouraged to learn on his return that there had been a drop in the number of deaths. Only 19 people had been dragged off and eaten in Jifiki's Wangingombe village since George had been gone; hardly good news but at least an improvement. Yet, the pressure for the reinstatement of Matamula as Head Spook by the residents of the Njombe district as a whole was growing. In fact, a deputation had gone over Jifiki's head to see the Paramount Chief about this, a rare event in rural African political structure. That the Paramount Chief, on government instructions, had refused to give them a hearing had not made them very happy. Clearly, George had better start cancelling out some more lions, and quickly.

Although he tried to buy it back, the new owner of the 9.3 mm double rifle refused to sell and Rushby was forced to buy a Westley Richards Mauser action .404 magazine rifle, a very fine caliber with excellent stopping power, but the bolt action not what George was used to with the double 9.3 mm. Deciding to use Mambego as his headquarters, hunting only with Alfani, he brought along a pocketful of candy for the pretty little girl who had given him the eggs. He was too late. She had been killed and eaten by the lions four days before.

The village was a wreck and the people like zombies in their grief and fear. Touched by the awful death of the child, Rushby set off with Alfani southward, looking for fresh spoor. On the third day, sleeping on the trail, they found it, the tracks of an entire pride of man-eaters. Following as quickly as silence would permit, the hunters pushed on through the thorn country until the rearguard came into view. As planned, Alfani and George fired simultaneously at different lions, the African's first slug wounding a lioness and the second missing cleanly. Rushby flattened his lion with the first shot, but it got up and had to be belted again, falling dead. George pumped two more bullets into the female Alfani had hit, killing her and saw the scout fire at a third cat at the edge of visibility, hitting it but not seriously. Although the men followed a long way, the blood trail quit and, as the wounded animal was able to keep up well with two other accompanying it, it was decided that it had only been creased. In any case, there were now two less of the Njombe man-eaters.

Following the killing of the lions with Alfani, Rushby lost a great deal of time in what was known as the Marshall-Cathles affair, a case of a man being killed by an elephant while involved with an ivory poaching plot, and George was unable to get back to the lions until February of 1947. While he had been tied up on other business, though, a lion had been speared to death by a young Hehe herdsman when the cat had tried to pinch a cow, although George doubted it was one of the man-eaters. Still, one unquestioned man-eater had been killed by a game scout at Mterengani village and two more *Simbas* of undetermined persuasion near Mawindsi. Most important, there had been an unquestionable decline in the numbers of human victims, Jifiki's area alone reporting a drop of about 50 percent! At last, Rushby started to feel that he was gaining the upper hand. Not only that, but he believed he had been able to figure out some sort of predictable pattern that the man-eaters used to determine their movements after a kill and on the way to another. By working with his map, George noticed that a sharp angle was inscribed as a retreat from an attack until it crossed a village, at which time the lions would kill and feed again. As the last deaths were a triple killing in the Malimzenga area, he guessed that the hamlet of Halali, some ten miles away, just might be the next stop.

Before first light, George was combing the *nyika* for some sign of the pride, hunting hard for six hours before returning to Halali in disgust. To his furious frustration, he learned that within one hour of his leaving to hunt, the whole pride had walked smack into the village, killed a man and eaten him within sight of the survivors. When George inspected the site, there was nothing left but the usual shard of brain pan, licked clean.

With the trail so hot the tracks were virtually smouldering, George had a quick pot of tea and made ready to follow up the pride. To his delight, there were two actual volunteers who wanted to go with him, obviously a refutation of the werelion belief; a grown man and a sixteen-year-old boy, both armed with spears. It took only a few hundred yards for Rushby to realize that the youth was one of the most gifted trackers he had ever seen, positively brilliant on the difficult spoor.

After about three hours of fast walking, the men reached a large area of what had been cultivated land, now grown over with dense brachystegia and combretum early in the fifteen-year reign of man-eating terror, making dangerous hunting conditions.

With the youngster fearlessly leading, following the tracks, George covered him with rifle ready, the older man behind him. Fifty yards into the snarls, a tiny sound could be heard, but what it was nobody knew. Stepping ahead of the boy, Rushby got down and began to crawl along, at twenty yards spotting a lioness not fifteen feet away the same instant she saw him. Before she could react, he whipped a .404 slug through her skull and rose to his feet to work the Mauser bolt action. Off to his right, about 30 yards away, another lioness leaped through an open space, catching a bullet too far back in the guts to be fatal as George snapped off a round by reflex. As she raised vocal murder, tearing and biting at the vegetation, George turned and was surprised to see that the boy was still with him although the older man had cut and run. From the sounds around him Rushby figured out that there were at least two other lions near the wounded lioness and frankly admitted to himself that he was scared spitless. To steady his nerves, he kept the rifle pointed toward the unseen lioness and lit his pipe to collect his wits. When he finished it, he took a deep breath and went in after the wounded man-eater, signing to the youth to stay put. Eyes rivetted on the place where the bush had been swaying, he literally inched forward for three steps, the blood crashing in his ears. There was the smallest sound filtering through the cover ahead.

The roar froze him with unspeakable terror for a heartbeat, the intensity of it slapping him from the left rear. Spinning, he stared directly into the malevolent, orange eyes of a tremendous male lion, his dark brown mane framing the face of death five feet away. The rifle was cocked and jammed against his hip, but he knew it was too late, the big cat just too close. For an eternity they locked eyes and wills, close enough for each to touch the other. And then, in a liquid flowing movement of massive muscle, the lion was gone into the cover. Rushby never did fire.

When he got his heart going again, George lit another pipe and called to the boy to come over. There was still the matter of the lioness. Together, they inspected the torn-up patch of earth she had savaged where she fell from the shot, and following the heavy blood trail, found the cat forty yards on. She was lying on her stomach facing the men and George instantly fed her two more bullets. She keeled over, her legs twitching, jaws snapping in her death throes. The boy tapped George on the shoulder, asking if he might spear her before she was completely dead.

Rushby stared at him a second and asked why.

"My father was the man who was eaten at the village," answered the lad without a blink.

"Stab her quickly. She is not yet dead."

The boy expertly drove the iron spear blade through the man-eater's heart. Emotionally exhausted, the men nearly collapsed and, on the return to the village, found the third member of their party very sensibly high in a treetop. Another week's hard hunting brought no further sign of either the big male that had scared ten years off Rushby's life or the second, unseen lion that had been with him on the spoor. On George's return to Mbeya for some of the usual bouts with paperwork, George decided to stop off at Iyayi when, driving past, he noticed Matamula outside his hut. He had not seen the witch doctor in months.

"Two more *dudu ya porini* have died," said George searching the man's face for any sign of reaction. "Ten are now finished."

Again, the studied casualness: "You have hunted very well," Matamula said evenly. "It will be very interesting to see the end of this matter."

Rather a curious retort, Rushby must have thought. Irked with himself at even speaking with the witch doctor, he pushed on to Mbeya.

As fate and the game department would have it, it would be nearly three months before George Rushby would get back to hunting the Njombe lions, mostly because of a terrific outbreak of genuine lion-man ritual killings further north in the Singida district, which, although certainly not a boring tale, would not be properly included here. When at last free of his other duties, George went straight from Singida back to the Njombe district, stopping at Wangingombe to see his good pal, Chief Jifiki. Rushby was delighted at Jifiki's news that Alfani had killed another definite maneater only ten days before and that the rate of deaths was so far down that there couldn't be more than a couple of man-eaters left. Also, they seemed to be acting differently than before, much more predictable in their behavior, their only stronghold seemingly around Jumbe Musa's village.

Relocating in Musa's village despite the usual animosity from the inhabitants who believed that hunters only antagonized the man-eaters, George and his scouts hunted for eight hard days before cutting the fresh spoor of two lions, by the track, one a very big male. George prayed that it might be the huge brown-maned chap that had scared him so badly in the scrub. Instructing Jumapili, the man with him, to shoot only the lioness, George swore to kill the male if he had a chance. Three hours later he was to have that chance.

It was almost an anticlimax. The lions were lying in the shade of a thorn tree, warming up to a bit of romantic dalliance when the male heard the men and stood up. Rushby instantly recognized him and stuck a .404 through his heart. Jumapili ventilated the female's lungs, killing her with a fast back-up shot. George walked up to the dead male and looked into the face that haunted his dreams. There was no doubt. It was over.

George Rushby reckoned he had never had a more pleasant duty than to wire the retiring Wenban-Smith of the end of the man-eating reign of the Njombe lions.

Settling in at Mbeya with his wife, Rushby returned to his normal duties of elephant control and the thousand other responsibilities of running a bush community. He was still trying to catch up in the beginning of June when a telegram arrived from Wenban-Smith:

"YOU'RE WRONG GEORGE. REGRET. WOMAN EATEN YESTERDAY AT MATIPU VILLAGE."

Rushby blew his vocal top. It had meant a great deal to him to be able to let Wenban-Smith leave with the Njombe man-eater matter closed on the books, not leftover for his replacement. And, now, with the new work he had implemented since believing the terror to be over, it wasn't until July 16th that he could get back to the Njombe district, pitching up that day at the normally dour village of Jumbe Musa. Upon his arrival, he wondered if he'd taken a wrong turn somewhere. The usually resentful people were primed for a blow-out celebration, brewing beer and slaughtering stock. As George quickly learned, the scouts he had sent had killed two lionesses, almost surely the last of their man-eating kind, just the previous day. They were numbers 14 and 15 of the definite man-eaters, not counting five others of questionable category killed outright and two more which had surely died of wounds and been lost. George drove off to Wangingombe to share the festivities with Jifiki, who had been through so much of the campaign with him.

After a few beers with the chief, George was ready to push off back to Mbeya when Jifiki rather awkwardly stopped him for a final word.

For a few minutes, they discussed George's years in Africa, his future with the game department and the campaign against the lions. And then, Jifiki said an interesting thing:

"Don't expect the people to ever believe that the ones you shot were the lions that were killing them."

Rushby blinked incredulously. "Why not?" he asked.

Jifiki was clearly uncomfortable, unbeliever that he was in black magic and Matamula's imagined part in the man-eating. Shuffling his feet, he told George that, when word got out of Wenban-Smith's leaving, a large local deputation was sent to the Paramount Chief demanding that Matamula be reinstated. During the period between when Wenban-Smith left and his replacement arrived, the decision would be up to the Paramount Chief and not the government. Politician that he was, the Paramount Chief agreed, removed Ulaya as subchief of Iyayi Village and replaced him with Matamula. The populace was delighted and the grapevine word had gone out that Matamula had called off his lions. There would be no more killings now that the witch doctor was returned to his proper place.

George Rushby slowly filled and lit his pipe, staring at Jifiki. Twice, he started to say something, but changed his mind. Without another word, he got behind the wheel of his vehicle and drove to Mbeya where his family waited. There have not been any recorded killings by man-eating lions in the Njombe district since.

Interesting place, Africa.

* * *

It's just as well that most covens of man-eating lions don't have the degree of extended success that the Njombe lions experienced or most of black Africa today would probably be nearly uninhabited! But, it doesn't take upward of a generation of *Simbas* killing a couple of thousand people to create the same all-pervading terror, albeit on a smaller scale, that I have encountered several times in neighboring central African countries during my years there as a professional hunter and erstwhile game control officer. Even one effective man-killer moonlighting part-time is quite sufficient to put entire districts into a collective funk. Three man-eating lions who were really starting to get the hang of their new line of work in the area surrounding the small village of Kalundi, about ten years ago, in Zambia's Northern Province when I was a game control officer were just about enough to shut down half the bloody province, so I have some idea what George Rushby and his stalwarts had on their hands. Here's a rough idea of what happened—and what *almost* happened. . . .

* * *

The sun of Zambia's Northern Province was fully up, a great incandescent globe that seared through the stand of mucassa trees at the edge of the safari camp, spearing the packed earth of the compound with brassy lances of light. Automatically, I swung my bare feet over the edge of the camp bed to the impala hide that lay on the dirt floor of the grass and sapling hut, momentarily confused as to where I was. The sudden movement awoke a stab of pain behind my eyes as the hangover hit me, my stomach full of hyperactive bats and the metallic alloy of tobacco and liquid indiscretion thick on my dry tongue. Carefully I eased back, scrabbling blindly for the aspirin bottle and popped a couple past my wall-to-wall nylon shag teeth, washing them down with the mug of now cold tea left for me by some saintly soul at what must have been dawn. Whoever it was must not have been present on the night before. . . .

Gradually, the pickaxes stopped prospecting the interior bone of my temples and I tried to reconstruct the previous evening, bits and pieces beginning to fall into place. I had pitched up about dusk at the safari camp of an old pal, professional hunter, and sometimes-mercenary, John Welford, who had just himself returned from depositing his last batch of clients at the company's bush airstrip—a pair of Texans he had been playing Rover Boys with for the past five weeks. By the look of the trophy racks by the skinners' quarters, it must have been a pretty good hunt, too.

Having been out in the "blue" on elephant control for quite a spell, I was looking forward to finding out if I remembered any English as well as tossing a few toddies with John, with whom I had done several safaris before joining the game department. I suppose that, since we both had a valid excuse for some celebration (he because of a 12-day break between safaris and I through simple anticipation of a night in a real camp bed and something other than mealie-meal mush and venison) it was not entirely illogical that things got a bit out of hand. I remembered the scotch and the three bottles of bright Cape wine with dinner and even the Boer brandy around the fire, but things sort of faded somewhere around the stingers. Must have been the altitude.

I wrapped a cotton *kikoy* around my waist and stepped to answer the polite rap on the *mopane* wood frame of the hut, swinging the door open on greased, leather hinges, rigged inside so the hyenas wouldn't gnaw them away. It was Chandiri, John's cook, with a mug of strong, sweet tea which I swilled down, feeling the stuff sandblast some of the taste of last night away. I grabbed the shaving kit and, spotting the kitchen *toto,* shouted across the camp to him, *"Faga lo manzi tshisa manje, yabolisa!"* and saw him scoot to fetch the hot water for my shower. After three buckets I started to revive and, even though the edges of my hands still tingled, at least most of the headache had wandered off in search of easier prey. I brushed my teeth, shaved with the care of a brain surgeon working on his first "big one" and went back to the spare hut to slip into fresh bush shorts and the epauletted shirt of department issue.

"Bwana Game," called Chandiri from the kitchen, *"lo skafu ena kona lugili manje!"* Hot dog, chow's on! I went over to the open-walled dining shelter where a big platter of scrambled eggs—the real ones from genuine chickens—was joined by another plate suffocating under its load of beefsteaks, each steaming and edged with glorious yellow fat. After a lean diet of impala and buffalo for weeks on end, that table couldn't have looked better to me if it was stacked with platinum ingots. John Welford came sauntering over from the skinners' area, wearing just bleached bush shorts and sandals, looking as if he'd been asleep for a week. The guy was absolutely hangover proof.

"Sakubona, Inkosi," he saluted me in Sindebele Zulu. The big Rhodesian looked at me owlishly. "Gawd," he said in a voice filled with awe. "You look absobloodyloutely terrible! I've got stuff over there pegged out that's in better shape than you." He scraped a minor mountain of eggs onto his plate and, as if using *banderillas,* harpooned a pair of steaks, all of which began to disappear at an alarming rate into his mustachioed mouth. Afraid he might be going for seconds, I quickly loaded up and finished breakfast with the standard African cure for overindulgence, an icy bottle of lager from the parrafin fridge. If there is a Heaven, it couldn't have

been more than a couple of hundred feet away. As I pushed back from the table, pondering the addition of another couple of eggs to the stack in my stomach, I noticed John looking past my shoulder, his dust-and-glare permanently bloodshot eyes narrowing perceptibly. By reflex, I turned around and saw the figure of a man trotting steadily toward us, his image distorted by the growing heat mirage over the low, grassy *dambo* on the far side of camp. As he drew nearer, I noticed he was carrying a split stick with a scrap of paper wedged in it. He pulled up at the perimeter of the safari camp and squatted, waiting expectantly for an audience. John and I went over to him, the Rhodesian calling for one of his men to bring a calabash of water. Without more than a brief greeting, John took the stick and unfolded the sheet of lined notebook paper. Looking over his shoulder, I read it:

> Dear Sirs and Gentlemens.
>   In this places is much bad deads from the lion eatings. her is much peoples dead. we beg and pray you very much you come kill the lion eatings very much. is three lion eatings. thank you for come qiklie very much.
>
> sined a mission boy of kalundi.

John and I both looked down at the man and then back at the pathetic plea. "That about cuts it," Welford said finally, switching to Fanagalo to tell the messenger to wait while we got our gear assembled. I called for my old Awiza gunbearer, Silent, who was encamped with Welford's staff, and he jogged over. I had been traveling in John's area with just the old man, putting pressure on a small gang of jumbo who had been raiding village granaries, and had left the rest of my small staff back at my control headquarters some hundred miles away. It had taken a lot longer than I had thought to sort out this group of six bulls, but I had finally managed to catch them red-trunked a couple of nights ago and killed two, leaving a lasting impression, I'm sure, on the survivors.

In twenty minutes we were ready to roll, my short wheelbase Land Rover packed with provisions, men, arms, and a few of the homier pleasures of Scotland plus extra flax jaw-sacks of water hanging from the supports of long-lost rear-view mirrors. I was mighty happy to have Welford aboard; he and his gunbearer, Karinga, were a fine team to have on your side if things got sticky. And, with a fistful of people-eating lions in the offing, I had a feeling things just might get sticky.

The hot, dry miles thumped and rumbled under the Dunlop "Safaris" as we followed the directions of the messenger, an understandably morose man who Silent had discovered to be the widower of one of the man-eaters' victims. He sported the dignified if unlikely name of Gladstone. I considered with Welford whether to question him as we rode along, but the annoyance of the wind whip and seeming reluctance of the man to converse beyond hand signals dissuaded us. We'd find out the score soon enough, anyway. . . .

The rich, classically amber light of the central African dry season afternoon was behind our backs when we reached the ramshackle village of Kalundi. More familiar with the area than I was, John told me it contained about 40 persons in its collection of mud-walled thatched *kaias,* hard on the northwest bank of the Luangwa River. Except for the fact, however, that its citizens were being eaten at a very steady rate by lions, there was nothing special to differentiate it from most bush villages in that part of Africa—desperately poor, its villagers ignorant of the most elementary forms of modern agriculture, and as generally filthy as an 11th Century English hamlet. I pulled up at the edge of the village clearing and turned off the ignition key. Shy of strangers under even the most social of conditions, nobody was in sight. Yet, there was a pervading, almost physical aura of terror in the air, almost a smell I could segregate from the eternal tribal dwelling odor of millet beer, smoke, stale urine, and badly cured biltong. In Kalundi, being in a tsetse fly area, there was not even the cackle of a half-starved hen or slat-ribbed dog's whimper to break the spookiness in the air. Welford stayed with the car, keeping his eyes open without seeming hostile as I walked over to the largest hut, by its size and position it was that of the headman. *"Hodi?"* I called in the universal KiSwahili vocal knock, a greeting that has spread through the hoary centuries to areas where KiSwahili was never even heard of. An English translation would mean, "Hello, the house!" yet it properly means a request to "come near." A muffled *"karibu"* or "come in" sounded through the walls. I saw the door sag as reinforcing sticks were drawn aside and swung inward on flimsy *ntambo* bark rope hinges. I stepped into the gloom and immediately wished I hadn't.

Slowly, my eyes became accustomed to the darkness of the hut. On a pallet of badly cured hides lay an old man, his thigh and shoulder swaddled in greasy, pus stained rags. A sag-breasted crone squatted next to him, staring at the ground with a vacant face. I stepped over and wordlessly touched the old man's forehead, a slab of hot, dry skin stretched like a drum head across his bony skull. A glance as well as another whiff of the wounds under the rags told me two things right off: First, he had been worked over by a lion, and quite a thorough one at that; second, he'd seen his last sunrise.

I left the hut and returned to tell Welford about the headman. He didn't seem to be in much pain so near death, but I sent Silent with water and some 50 mg Demerols to help grease the tracks a bit for him. Reckoning it was about time to find out what had been happening, I blew the Rover's horn in three long blasts and shouted for everybody to come out. Hesitantly, doors opened. Perhaps 30 people

(55)

edged closer, seeming a bit relieved at seeing John's Holland & Holland double rifle, a .500/465 Nitro Express, an extra pair of long cartridges dully gleaming from between the fingers of his left hand. Everybody seemed shy, staring at their feet and fidgeting. All but one.

He was tall for an Awiza, probably in his twenties, although I make no claim to be able to accurately judge the age of bush Africans. Certainly, he was fit. He carried two spears and the standard tomahawk hatchet thrust through an old leather belt at his waist. He had been given the Christian name of Aaron.

Two weeks ago, he told us by indicating the fullness of the moon with his hands, two women—one the wife of Gladstone—had gone in the late afternoon to carry water from the river. Gladstone's wife had already filled her clay jug and stood some yards away from the other woman, waiting for her to fill hers. Suddenly, there was a tremendous snarl and a huge, pale-maned lion charged from a clump of riverine bush. He hit the waiting woman, paralyzed with fright, flat on, and with a swift lunge grabbed her by the head, driving long fangs deep through her skull and into her brain, killing her instantly.

Two days later, continued Aaron, since there had been no more sign of the lions, a man had gone to check his light *kwahlie* or partridge snares and had not come back that evening. The next day, his right hand was found along with a homemade coil of bark fiber snaring cord and his hatchet in a heavy snarl of cover no more than a mile from the village.

The following evening, just as the fires were burning low, a sleeping woman was dragged right through her hut door, torn off by one of the man-eaters and carried off, screaming. Aaron said that she had screamed for a very long time, then abruptly stopped. Her husband did not even realize she had been taken until he sobered up from the death-like sleep that the powerful *tshwala* millet or corn beer produces. Half mad with grief and guilt, he quickly took his spear and began tracking the lions. Nothing, no trace at all, had ever been found of him.

Of course, four nights ago the *kaia* of the now-dying headman had been savagely attacked by the three lions who, failing to breach the roof, finally ripped off the front door. The male grabbed the man by the shoulder and dragged him outside where one of his girlfriends nailed him by the thigh. Together, they started to carry him off into the darkness. Shocked awake, the headman's son, a powerful young man who had been staying in a different village until the attacks, came running from his hut naked, brandishing his spear. With incredible bravery he charged straight at the lions. The cats stopped, confused, but did not drop the old man until the son whipped his spear at the female holding his father by the leg. The razor-whetted iron flicked through the moonlight and sliced bloodily—but harmlessly—through the skin and thin meat over her shoulders. She roared and jumped convulsively, the shaft of the spear still caught in the wound, cartwheeling free against the night sky. Then, with twin roars, both she and the lion dropped the headman's mangled form and, as one, swarmed the son from only a few yards. The crushing of his bones could be heard through the village, said Aaron.

Aaron casually pointed to the son's spear, still firmly embedded point first in a pole of the son's own hut by the lioness' convulsive leap. No one would dream of touching it. Clearly, it was *Mbojo,* black magic thrown by the lioness.

After Aaron's briefing, I gave out a pair of cigarette packets to be distributed among the men and gave instructions for the burial of the headman. When he was interred and thorn bushes planted above him, I organized a party to cut more thorn, the nastiest, spikiest available to make a series of *zaribas*—as they're called up north—barriers around the huts to at least slow down the lions when and if they returned. I gave Silent my shotgun to cover the thorn gathering party and John lent his to Karinga while I broke out my Evans .470 Nitro Express double rifle. We would stay in the village on the chance that the cats might show up while the men were gone.

Since four days had passed since the last attack, it would be dumb luck to blunder into fresh spoor even with tomorrow's daylight. The best thing would be to stay put here at the village in ambush, since the lions had learned that each visit here had produced a kill. And, after four days of fasting they would be getting a bit peckish, very possibly giving us a chance tonight. My personal choice of a vantage point would be a *machan* or tree platform that would give a good field of fire and visibility over the village compound, yet keep me out of reach of surprise attack should they spot me before I saw them.

It was pretty well dark by the time the village was prepared, the sun sliding down the cloudless, dry season sky to disappear like dying fireworks over the distant mountains.

I settled down and opened a box of five rounds of the British Kynoch ammo, big soft-point slugs that looked as nasty as a nest of Gaboon vipers, their blunt, blue noses glinting darkly in the dim light. I pushed the lever and tonked a pair of the cordite rounds into their chambers and closed the action. With the rifle across my knees, I then fed five buckshot loads into the Model 12 and topped them off with two Brenneke rifled slugs, the last in the pump gun's chamber.

Then, with a frozen stab of realization I caught definite motion at the rear of the hut to my left front, a half-misty form gliding through the gloom. Adrenalin pumped through my system as through a fire hose, my pulse hammering in my ears, yet my hearing heightened. Another movement and disappointment welled through me as a tremendous hyena hunched out into the moonlight, swinging his muscular head back and forth as he shuffled across the compound. I eased a shotshell out of my pocket and tossed it at him. As it plunked off his ribs with a hollow thunk, he scuttled like some obscene crab back into the bush.

The booming twin roars of both barrels of John Welford's .500/.465 fired almost simultaneously very nearly scared me out of the tree in an involuntary leap, the muzzle flashes of 146 grains of cordite flaring the shadows at the other end of the village like heat lightning. I heard a curse and several growls. Heavy movement. Something flashed between two of the huts and into the thick bush, then another tawny, silver-plated streak flickered like a wraith through the thin moonlight. The third time, I was ready. I saw the long form clear the edge of a hut, streaking for the same spot as the first two lions and swung the .470 ahead of its path. When the bead was passing the tip of its nose I fired the first barrel and, at 40 yards, heard the heavy *whunk* of a big slug smashing meat; then thrashing. Semiblinded by the muzzle flash of my own piece, I fired the second barrel in a snap shot at the spot where the lion had disappeared, then snatched up the shotgun and hosed the area. I could hear the Brennekes whining off the dry ground to rattle through the trees while the buckshot chewed up the foliage, clipping twigs and slapping into the dirt. I reloaded the Model 12 with #1 buck "baby mags," restoked the Evans, and called for Welford to cover my matchless body while I got the hell out of that tree. If there was an answer, I didn't hear it.

He was standing outside the hut trying to get a cigarette lit, and not doing very well. His face was chalky even for moonlight, and I realized with a jolt that the front of his bush jacket was torn and dripping blood heavily. He leaned back against the hut wall as I came up, looking at the gore on his fingers in a very strange, detached way. "Jaysus," he kept repeating as he fingered the slippery stuff, "will you look at *that*!" I got him to take off the tatters of the jacket and took a look at his chest. Four bloody furrows sliced down it at a slight angle from a paw stroke, leaving a nicely patterned set of eight-inch gashes, shallow but mean, oozing blood fairly heavily from between white, puckered wound edges. For sure, from now on he would be the perfect macho image any time he got into a swim suit. (Where'd you get that, big boy? Oh, just a man-eating lion.)

"What happened?" I asked him quietly. Silent and Karinga came up with their spears, Silent lugging the *bokis ga muti* or medicine chest as well as a bottle of scotch from the bedrolls. Welford took a short blast, then another. I got his cigarette lit as well as one of my own and some color began to come back under his tanned cheeks.

"Go to hell if I know," he said shakily. "I saw this one bastard walk out of the bush, plain as you please, and laid the sights on her. She was setting me up, keeping my attention! Then that big male swine sneaked around the hut and reached in for me, cool as ice!" Bare chested, Welford sat watching me inspect the claw marks in the light of the flashlight.

We could go on from there and spoor up the other two of the trio if John felt up to it. Up to it? Welford explained with great, if somewhat off-color clarity, that any lion that got a piece of him would pay for his indiscretion with portions of the extremity of his elimination tract. Leaving Silent and Karinga to split what little was left of night guard, we turned in.

Dawn bloodied the east when Welford, sore but ready to go, joined me in biltong, saltines, and tea while checking over our rifles. We decided to stick with the doubles today because of their combination of stopping power in a charge and their greater range over the shotguns. It was cold as a crypt in the early light, dressed as we were for stalking: shorts with no underwear to chafe when it got hotter and sweatier, sockless sneakers so sharp grass seeds would not stick, and sleeveless bush jackets to avoid as much as possible the scrape of cloth and foilage. Silent and Karinga were shivering in their thin shirts at the place we picked up the blood trail where I had hit the big lioness the night before. We began to follow it into the heavy, riverine thornbush.

If there is a more potentially dangerous situation in hunting big game than following confirmed, wounded man-eating lions through heavy cover on their terms, I must admit it's missed my direct attention. I wasn't all that bothered by the lioness I had hit; I knew what one of those big, blue-nosed .481 diameter cannonballs would do, and the bullet sound had told me she would be lying dead just about ahead. It was her companions who could be and probably were lying in ambush under just about any chunk of shadowy bush you wanted to pick, waiting silently, unseen until we were on top of them. Surprise! And they would be on us in the characteristic tawny blur of low rush, impossible to stop from point-blank quarters, even if you had been living the clean life. With John and I looking as if were were walking in to take the flush of a pointed ruffed grouse, Silent and Karinga worked a few feet ahead, wordlessly pointing with grass stems or spear blades at the dry splotches of lung blood here and there.

Silent held up his hand and, not turning his head, slipped behind me, clearing a field of fire as the spoor ran into a big clump of bushy, dense combretum shrub, an impenetrable colony of bushes that could hide an elephant herd, let alone three lions. The old man had done his job. Now it was our turn. I thumbed the safety off the Evans and inched toward the clump, John just to my left rear so I would not block his fire with my body. I could clearly smell the acrid spray of lion urine in the damp air under the bushes and spotted the pug-marks of the other two lions as clearly in the soft dirt as if cast in plaster of Paris. Never having been especially talented at mathematics, it was just dawning on me that there were three lions and I had two barrels; which might add up to minus one. Me! Still, whispering a small prayer that Walt Disney and Cleveland Amory were right after all, I pushed aside the first of the green screen with the muzzles, trying to see something in the dank, mottled murk. My heart jumped as I saw a patch of pale hair in a sliver of light about 15 feet away (which is almost exactly 15 yards too close) and instantly fired the .470, the blast filling the tangles with the delicious reek of cordite fumes. I braced myself for something to happen, the express sights still locked on the patch of hair. Nothing did. I quickly broke the double's action and the ejector pinged the empty brass over my right shoulder as I beat its plunk to the ground with the closing of the breech over a fresh cartridge. I crawled over to the dead lioness. The bullet had not been necessary; she lay on her side, her tongue nearly bitten through in her death throes, dead for at least six hours. There was also about 15 pounds of meat missing from her chest where her pals had fed upon her, not unusual with lions.

I looked over my shoulder at John and he waved Silent and Karinga forward to help me pull the carcass into the open, covering the operation with his Holland with gratifying care. Back in the sunlight I saw that the shot the previous evening had taken her just where I thought it had, through the air conditioning, a bit high and back from where I would have placed it precisely, but the bullet had sure made a fair imitation of the Lincoln tunnel through her upper chest. I also saw I'd creased her with one of the Brennekes and there were five buckshot pellets embedded, but not deeply enough at that range to have any effect. John and I inspected the crusted-over spear wound through the loose skin over the shoulders from the throw by the headman's son. Pity. A foot lower and he would have had her, instead of the other way around. . . .

I squatted at one side of the dead lioness and offered John and our men a smoke. As I reached over the carcass to hand one to Welford, a small movement caught the corner of my eye behind him. "Watch it!" I yelled. "Behind you!" I snatched for the Evans as John swung around in a crouch, his .500/.465 ready. There was a whisper of bush and a ghost of movement. Welford fired twice, but I held back in case of a charge catching us both with empty rifles. I now covered him as he reloaded the 480-grain rounds. Together, we worked our way 30 yards back into the cover.

There, I saw with an involuntary shudder, were the pug-marks of the other two lions, still so fresh that if it had been cold enough they would have been steaming. The man-eaters had been *that* close! Despite my firing at the dead carcass, they had stayed where they were, less than a hundred feet away, watching, waiting for their chance when we dropped our guard. I believe that the motion that caught my eye was the male digging his hind feet into the soft earth for purchase just before he would launch himself. John shook his head in disbelief and we sat down, four around a tree, back to back and ready.

"Might as well give those mothers time to settle down before we take it up again," said John, who did not have to make the suggestion twice. The dead cat was pretty torn up, making the skin worthless, but I wanted the skull for a memento and told Silent to cut off the head. I'd set it on an ant heap after skinning and within a day, it should be clean as porcelain. He finished the decapitation, slicing out the floating collar or "lucky" bones which I dropped into a pocket, and we buried the skull off the path to pick up on the way back.

We snubbed out our second butts and took up the trail again, not all that convinced who was hunting whom, through a nightmare of thorn and thickets as impenetrable as accordion wire, along which each step might bring a charge from almost underfoot. The spoor went almost in a straight line for about three miles, the lions not even stopping to listen for us, which was damned unusual compared with normal lions.

It was nearly noon when I saw it, the soft curse from John's lips matching my own. It was not "shuckin's." Ahead, looming through the wavy heat mirage of dry bush towered a huge *kopje,* a great 10-story high pile of gigantic boulders stacked atop each other unknown millions of years ago by some natural action, a honeycomb of natural caves, and black passages offering a fortress for the lions that stacked the odds against us like shaky dominos.

When we reached the base, we walked around the stone mountain, examining the rocky openings that led deep into darkness. It could hold, in addition to a pair of man-eating lions, cobras and mambas that love to hunt in the rocks for hyrax and small rodents. The *kopje* might also be home to a leopard. Well, we still had to give it a try or people would be dying again in in the villages. While Silent and Karinga cut bright burning mopane for torches, John and I spoored around, trying to decide where to start. We had lost the spoor on the rocky approaches to the stone hill, but one large cave seemed a good choice because there was ancient Bushman art, rock paintings of giraffes, hands, and running, elongated red men with bows near the entrance, which led us to believe that it went well back into the hill. Torches flaming in the blacks' hands to leave ours free to shoot, we started into the maw of the *kopje.*

It didn't take 10 feet to know that we had the right cave. The acrid, wet-cat odor of lions permeated the place, hair filled dung balls and bone scraps thick underfoot. Silent stopped suddenly and, holding his torch low, bent and picked up a spear, the iron blade still bright and unrusted. So this was where the man whose wife was killed while he lay in a drunken stupor had met his fate. God, but what courage he must have had to track the man-eaters all the way to their lair, a lone man and a single spear against three of the deadliest of predators! Ten yards deeper into the cave we found a burned-out torch and, scattered like grisly dice, fragments of crushed jawbone with some human teeth still held in place by dry scraps of half-mummified gums.

The growl was terrifying, coming from everywhere and nowhere at once in the confines of the large cave walls. I threw myself backwards as a huge, golden shape cut through the torchlight from above and behind, grazing my head to land with a heavy, cushioned sound on the rock floor a few feet ahead with a roar that paralyzed me like a punch to the solar plexus. Somehow, I fired the rifle at the blob, the twin shots licking long tongues of blinding orange muzzle flame that blended with the rocket of pain that shot through my right hand and forefinger. Then there was a universe of impossible sound and impact as something slammed into my back like a freight car, sending the rifle slamming into my chest and the stone floor rising to meet my head with a bright explosion of flashbulbs.

Gradually, I was aware of a terrible weight on my back and a flood of hot, sticky wetness soaking warmly through my bush jacket. My God, I thought groggily, the bloody lion is on me! I struggled ferociously for a moment, then felt the great weight rolled off. I sensed movement and could see the torches at some great height above me, John's voice coming down as if he was speaking through a drain pipe a long way away. Then, the world went black again.

I awoke 10 minutes later, John later told me. I felt like I had done a half-gainer in pike position from the high board into an empty pool as I lay there, the Rhodesian tying his handkerchief around my head. Slowly, things began to swim back into focus and I realized that there were two dead lions lying in tan puddles of gore alongside me. "What the hell happened," I managed to croak, violently willing down the urge to vomit. My head pulsed like somebody had stuck a blasting cap up my nose. I explored the big, square laceration on my temple, a square of cube steak perched on a lump like a vulture's egg where I had hit the cave floor.

"The lioness there," Ian pushed the body with a sneakered foot, "hit you dead on just after the male missed and you hammered him. I saw where he came from and was ready for her." Ian raised his torch and showed me a narrow ledge eight feet over the cave's floor where the lions had lain in wait. "Still, it was one flaming lucky shot, old boy. Caught her absolutely smack in the ear by the luckiest fluke, so even though she did hit you—and rather hard, I'd say—she'd had it." Good Garden Peas! If that's what it felt like to be hit by a dead piece of 400 pounds of lion, what would a live one be like?

Welford lit me a smoke and washed off the blood from my face with one of the flax water bags. Still running on two cylinders, I was helped out of the cave and back into the sunshine, feeling the lioness' blood drying, cementing the bush jacket fabric to my back. I sucked at the gash the rifle lever had cut in my thumb webbing and held it, looking rather stupid, I'm sure, over my head to reduce the flow of blood. Although we soaked it as best we could, when Silent pulled—or better said, ripped—the jacket free, it took most of the hair off my back and I plead guilty to the commission of one most unprofessional bellow of pain. A few minutes later, the men dragged out the two lions, first the male and then the old female. I noted with relief that, having pulled off both barrels in the same motion, the first had caught the lion in the chest, the recoil happily kicking the second barrel into position that the slug exited at the precise instant it was lined up with the bridge of his nose. If I'd missed the brain at that range, even a heart shot would have left him enough steam to spread me all over that cave before his brain ran dry of oxygen. Welford's shot was precisely as represented. There was no off-side of the lioness' head. I suggested he might consider taking up skeet shooting. "Do they bite back?" he asked.

"There, me dear old pal and Bwana," I said pointing at the two corpses, "but for The Grace, go thee and me. . . ."

Welford looked for a few seconds, cogitating, then shrugged. He motioned over Karinga, who carried John's shooting bag, and dug down into it, birthing a bottle of Haig like a magician drawing a rabbit from a hat. He cut the foil with a big, dirty thumbnail, pulled out the cork and handed it to me.

"Yeah," he said slowly. "But it isn't." I took a long swig and passed him the warm bottle.

Leon Parson

CHAPTER THREE

CROCODILES

Carmine bee-eaters are flickering over the river in the early morning sunlight, their shameless red and short circuit electric blue glistening with the purest color of tumbling, twisting gems as the young woman approaches the low bank of the Munyamadzi. The lifting lances of central African sun are starting to warm her bare, black shoulders as she walks gracefully to the shore, an empty calabash balanced lightly on her head, her mind full of her new husband. True, she is only a second or junior wife, but her superior is friendly and happy to have her share the long days of hard work in the fields, cutting steel-hard mopane firewood and carrying water. Already, the older woman's rebuke for permitting the water to get low at the *kaia* is forgotten in the glory of morning coolness and the flight of the gleaming *zinyoni*, swooping and chandelling, seeming to almost touch the water before sheering off with their insect prey. They are streamlined streaks of darting feathered brightness against the sere, dead grass of the far bank.

Singing softly under her breath, the girl removes the dried gourd from atop her head and, feeling the crispness of night-cold sand between her toes, steps into the water at the edge of a bar. She slowly wades out a few yards toward deeper water where she may fill her container fully without drawing in bottom sediment. Dunking the worn, brown vessel, she watches the river slide smoothly into it, filling it almost to the brim. She grips it carefully with both hands, about to make the single,

(65)

fluid motion to lift it back onto her head. But, it is too late. She has time to see the last ooze of stealthy movement and the burst of foaming speed before long, thick spikes of teeth slam together over her wrist and arm. She has time to look into the slit, flat, cat-like eyes of living death before the irresistible, numbing wrench pulls her flat, choking her with water. Terror courses through her 14-year-old body, but she is gagging too badly to scream as her arm is broken and dislocated at the shoulder. Her free, groping hand feels the hard armor of the creature's back as the current of the Munyamadzi closes over her. She will live another 45 awful seconds, her brain still working in unspeakable horror as she fights the relentless grip. Then, with a sigh that releases a string of wobbling bubbles in a silver chain to the surface above, unconsciousness yields to death.

On the bottom, patiently holding his prey, *Ngwenya* simply waits until all resistance is over. Satisfied, he swims off easily with his kill to a deep, quiet eddy where he may feed at his pleasure, ever more easily as the corpse decomposes. Above, an inverted calabash bobs and floats down to the junction with the Luangwa. A swooping bee-eater flashes briefly over it and then flies back to his mud dwelling in the wall of a high bank up river. A family of drinking elephants spray themselves with water, and a rhino bumbles blindly back to a shady thicket. It is just another African day.

Word reached me at my small game control camp nearby the day after the disappearance of the Headman's new wife. From the description of the presumed events, the only evidence being the girl's tracks leading to the water and not returning, there was small doubt in anybody's mind that she had been taken by a croc. She had been, in fact, the third victim in a year from this same village. Despite the protests of the elders, this was one matter which left little for me to do. I couldn't very well just start shooting each of the thousands of crocs that swarmed in the Munyamadzi in hope of getting the particular one that killed the girl, even if he was in actuality the same one that had eaten the other two women.

True, the past two attacks had been witnessed, but one croc looks much like another but for size, and this was just another 12-footer. I refused to even visit the place, knowing it would be a complete waste of time.

Often, I had suggested that a protective cage of stakes driven into the bank and bottom of the river would provide a safe place to draw water, but, as is so classically African, nobody ever went to the trouble. Time and again, a woman would be caught in the exact same spot by a croc as was a friend or relative a short time ago. They never learned and, despite what I have always felt was a good understanding of the native mind, this sense of fatalism continues to baffle me to this day.

The Headman would find a new second wife, maybe even younger and fatter. The men and women would continue to get motherless drunk on *tshwala* brew any time they could. Sooner or later somebody else would get eaten from the same stretch of river, just as they probably had been ever since these people had been chased up here by the AmaNdebele Zulu invasion of a couple of hundred years ago. Also, the sun would very likely come up tomorrow. Cynical? Maybe. Certainly accurate. Crocs will always be crocs, and bush Africans will always be bush Africans. So long as the two coexist, the first will always eat the second, and there is nothing you, I or anybody else can do about it.

Let me try to tell you the way of things in rural Africa. Eleven years ago, at the very moment the American Eagle was settling on the lunar surface, I was busy shooting 22 elephants as a cropping officer. It's hard to accept the incongruity of man's oldest profession (on the logic that the first Lady of the Evening was paid with meat killed by the first hunter) being conducted at the same instant his dreams of space conquest became reality. It's an example of almost the total range of human activity and experience still existing side by side in time. So perhaps you can better grasp the fact that even now, in the 1980's, thousands of people are caught, killed and eaten each year by the African Nile crocodile. Crazy? An exaggeration? I only wish it was.

The crocodile and his extended family are available in a dazzling array of styles and body types. In southeast Asia and the nearby Pacific islands, the wonderfully efficient salt water crocodile will eat you with unbounded enthusiasm. In parts of the Indian subcontinent, the "mugger" croc—cute name, what?—may pull you down to an astonishingly unpleasant end. Since the ban on killing Florida alligators, the death and attack rates have soared because the 'gator has lost much of his respect for his old enemies, thee and me, although we'll get into this in more detail later in this book. Still, the real overachiever in the people-purloining department is the African Nile crocodile (the same critter whether found in the Nile or your bathtub in South Africa) which may accurately be summed up as that land's only carnivore that will cheerfully kill and eat you every time he gets the chance.

Now, wait a minute. I know that's not what Lorne Greene told you last night on "Last of the Wild." Everybody knows the croc is endangered. So how could there be enough left to eat "thousands" of Africans annually? That's what those wonderful folks who can't wait to bring you the Worldwide Anti-Hunting and Firearm Confiscation Act of 1984 have to say on the subject. You know them, the same armchair self-sacrificers of *your* rights who got the leopard on the endangered list despite the reports of their own field studies clearly showing the status of the leopard so well populated that their own research biologist recommended the possibility of commercial hunting of them. Well, choose for yourself. But, my money's going to be on the chaps with the big teeth the next time you try your backstroke up the Zambezi or most other semitropical African rivers crawling with "endangered" crocs.

It is the very nature of the crocodile that makes him such a successful life form which, in all practicality, has had as little need to change over the past millions of years as has the great white shark. Actually, the two do share the common characteristic of happily including man in their well-balanced diets whenever the chance is afforded. Yet, where both are really simple eating machines, the great white is a mindless processor of protein while the croc is a master hunter and stalker, patient as death, as cunning as a weasel. And, I mean any big croc on a general basis. Which creates the rub if you're a game control type like me.

An outbreak of man-eating among lions or leopards can always be traced to a particular individual or small group which can then be hunted down and destroyed. The same with a Cape buff or elephant gone antisocial. But, with the croc, it's almost impossible to hang the case on a particular animal where reasonable numbers of the species congregate. Of course, to every so-called rule, there's an exception. I have written previously of the tremendous croc killed by a safari client of mine, Paul Mason, after the 15-foot-plus lizard killed and ate a woman right below my safari camp on a lagoon along the Luangwa some years back. In this case, we both saw the croc as he dragged the woman off and his give away was his unusual size. After days of hunting him, we finally caught him feeding on the carcass of a dead hippo we had shot for bait, and Paul atomized his skull with a .404 slug.

My old friend, Swiss-born professional Karl Luthy and his client, a Colonel Dow, were able to kill the croc that ate Peace Corps volunteer and Cornell graduate William Olson in Ethiopia's Baro River in 1966 mostly because he was the only large croc they saw in the area. It was taped at a shade over 13 feet.

In the tri-country central African areas of Zambia, Botswana and Rhodesia, crocs were, and to the best of my knowledge, still are on the general game license. Having operated as a professional hunter in all these places, it's fair to say I've been in on the Last Rites of quite a lot of crocs. Sport hunting crocs in daylight, as opposed to hide hunting with a night light from a boat with a gaffer, differs quite a bit from most Afro safari fare. It is not nearly so easy as those who have seen their *Ngwenyas* half-tame, loafing around the rivers of protected national parks would tell you. Crocs may be deadly when they are hunting, but are reversely shy when found in the only circumstances in which they know they are vulnerable—while regulating their body temperatures by sunning on banks and sandbars. Not only are they immensely wary, but the various ox-peckers, tick-birds and plovers that eat the vermin and leeches from their hides and even in their mouths are also very nervous, making the stalking of a big croc rather like trying to sneak up on a flock of hypertensive Canada geese. Sure, a decent shot can brain a swimming or floating croc in the water, but to what end? The carcass will sink like a pair of cement sneakers and either be eaten by the others or only rise when bloated and the skin spoiled. So croc shooting is much like American varmint hunting, a matter of careful stalking and accurate, long range shooting with a well-tuned rifle.

Although I am not a fan of the ultra-velocity class of hunting calibers because of my continual experience of having their bullets break up or badly deflect in our heavy bush, they are perfect for hunting crocodiles. I used to keep a super accurate little .243 Winchester with an 8X scope just for my clients' use on crocs. Of course, they had to zero in for themselves since no man can sight in another's rifle for him. However, once on the money, it was dynamite croc medicine, especially in 100-grain lead doses.

The only practical shot for the crocodile is to the brain, and that from a rather flat side angle. With the advantage of height on a basking croc, either a front or rear angle is possible, but at the same level, from head or tail-on, the body protects the skull perfectly. In any case, the brain shot is the only one that will practically anchor him then and there, although this is not a rule. I have seen crocs do some amazing things with nothing left north of their eyebrows—were they to have eyebrows. Even with their brains scrambled like eggs at a brunch, they can continue vestigial, reflex action for a disconcertingly long time. With a perfect brain shot, the tail will invariably explode into a mad flurry of action. If facing the water, this may propel the croc into the current, and he'll be gone. Forget it. Any animal wounded and lost or, in this case, dead and lost, counts against your license. If you are lucky enough to nail him down, chances are most of your crew will have to sit on him to keep him still for skinning.

As a collector of old hunting books, particularly Africana, I am constantly fascinated by the—to put it as kindly as possible—diversity of commentary written about crocodiles by what are, without question, some very experienced hands at African hunting. There is probably no animal on that continent that seems to attract the almost migratory flights of imagination from otherwise reasonable scribes (unless it be the cobras and mambas) than the croc. Of course, one contributing, if not mitigating, factor is the undeniable fact that *Br'er Ngwenya* does kill and eat more people in its broad range than any other species of human preying creatures. Among the carnivores, nothing approaches the crocs as an extended group of processors of living human meat, although few westerners realize that the hippo is the champion man-killer of Africa, far exceeding the inroads chewed by lions, leopards or hyenas, or the spectacular but numerically infrequent fatalities by elephant, buffalo or rhino. This is not to say that the preceding species don't slip plenty of people the big, bad news; it's just that the bad tempered ubiquitous hippo does it more frequently and, apparently, with more technique. The only reason he's not looming as a chapter somewhere in this book is that although he'll chew a man or woman into taco filling, he won't swallow you.

Among the more interesting aspects of crocs that I have enjoyed researching through the delicious tobacco-and-old-paper odor of the ancient books that gleam mellowly with faded, friendly gold-embossed spines from my shelves, clustered around my typewriter like the spirits of dear, old friends, is the matter of the croc's top size. Now, I have a personal "truth-test" for African writers that centers around crocs, and thus far it seems accurate: If the biggest croc mentioned in that particular writer's book is eighteen feet or under, and got away, the odds are that you can believe the vast majority of what he tells you about other species of which he writes. But, if he comes in with anything over 20 feet, he is automatically suspect unless having otherwise impeccable credentials, such as unrefutable proof of blood relationship to the Holy Ghost.

As might be expected, I find I partially fall into this category myself! In my first book, *DEATH IN THE LONG GRASS*, I smugly stated that the biggest (longest) African croc ever actually measured was one of 19 feet 9 inches, killed in the Semliki River, which forms the border between Uganda and Congo, and was purchased by the Marketing Corporation of Uganda in 1953. This is without question the same animal referred to by Peter Beard and Alistair Graham in their *EYELIDS OF MORNING* as being killed in 1952 and measuring 19 feet 6 inches. I am, after five years, no longer certain whose research was the more accurate, but for one year and three inches, I'm prepared to concede. However, this whopper may by no means be the largest measured Nile croc.

The problem lies with the fact that, as not being considered a game animal (although he can be extremely sporting to stalk) and his statistics not reposited in such heady records as those heretofore kept by Messrs. Rowland Ward of London as for other species, the "record" rather depends upon whom one chooses to believe. Here are some of the reports; see what you think.

The Duke of Mecklenburg is credited by two sources to have killed a tremendous croc during the teens of this century, although the dates vary slightly as do the measurements. It was listed by C.A.W. Guggisberg, (*CROCODILES*) as having been shot before the Great War, at Mwanza, 110 km. east of Emin Pasha Gulf in East Africa and measured, translated from the metric 6.60 meters, as 21.65 feet. The notes of the last King of England, expanded by Patrick R. Chalmers into the book *SPORT & TRAVEL IN EAST AFRICA* (covering the former Prince of Wales's safari in 1928 and 1930) listed Mecklenburg's toothsome beastie as being killed in 1915 and measuring 21 feet 6 inches.

The divergance of many reports is probably due, at least as far as actual measurement is concerned, to the method employed, i.e., whether the croc was still wearing the skin, whether the measurement was "between the pegs" (a straight line between stakes driven into the ground at nose and tail-tip) or over body curves, as a variable.

Another interesting report which is reasonably contemporary is from 1949, when a 21.98 footer was shot by professional hunter Erich Novotny, also in the Emin Pasha Gulf of Lake Victoria, an area which even nowadays must not sell much water skiing equipment. Also, a hunter by the name of Douglas Jones notes a measurement of 21 feet 11 inches from Somalia. No, I never heard of Jones, either.

If we want to gently probe the ranks of the really heavyweight division, we find some pretty reliable sources to quote. Mary Kingsley, a portrayal of whom many of you may have seen as the heroine of one segment of the excellent docudrama series which appeared on public television, *"Ten Who Dared,"* was the Victorian lady explorer who, despite being unaccompanied but for "indigenous people" probed the depths of West Africa alone. She writes of having measured a croc of 22 feet 3 inches and comments additionally that anybody reading the story back in Blighty who is not inclined to believe it pretty well deserves to be in England in the first place! Another incredible croc, killed in Lake Kioga by a Captain Riddick in 1916 was measured at a hair over 26 feet as was another one reported by the highly reliable and objective German writer and traveler, Hans Besser, who killed it in 1903 on the Mbaka River in what is today Malawi. Neither of these gentlemen gives the method of measurement as to between the pegs or over the curves. Tell you what; I don't care which measurement method was used, I wouldn't share a damp sponge with either of those babies!

So, what's the safe answer, yet staying within the realm of reason? Take your choice, but I wouldn't lose any sleep believing that there have probably been a couple of African crocs killed and measured by reliable sources (as well as many whose death was never recorded, or lost) who would have measured 24 feet between the pegs. Not many, for sure, but some seem within the boundaries of probability.

Right, chaps, those are some of the premier candidates for Godzilla of the Year that were, at least theoretically, measured after death. As might be expected, estimates of crocs *seen* by a bevy of explorers, hunters, missionaries and other traditional unreliables can really be mind-boggling. Some witnesses, though, had the grace to admit that they had greatly overestimated the size of a large croc after later killing it.

Before giving a couple of these Honest Abes their due, it's reasonable to grant that full visibility of the length of a croc is normally possible only when the animal is on land. When sunning, a croc tends to be seen upon sand or mud bars which are usually without any outstanding comparable feature for measurement reference, with the frequent exception of other crocs, who are themselves without a common known size reference.

A good example of this problem is exemplified by the observations of one of the pioneer animal photographers of Africa, Cherry Kearton, who once saw a tremendous croc laboriously crawl up onto a sandbank of the Semliki, which we know does produce some huge *Ngwenyas*. Kearton pronounced it to be ". . . not an inch less than twenty-seven feet." As he tells in *LAND OF THE LION,* (1929), it was in the company of several "ordinary" crocs of 12 to 14 feet—in my opinion and experience, this was a substantial exaggeration as to the average size croc found anywhere—except attempting to eat Tokyo in a Japanese movie—and created sufficient visual impact for Kearton to dislike crocs his whole life long. He didn't shoot it with a rifle, and the picture he took does make it look large, but only in relation to those near it, but I doubt they were anything like a dozen feet plus in length.

Colonel J. Stevenson-Hamilton, who certainly should have known his crocs after the years he was warden of the Kruger National Park, remarked on shooting a croc once that he was certain would measure 18 feet. When he recovered the body it taped out at just four feet less. That's a mighty big difference, but still a good-sized crocodile.

I had an experience similar to this myself in, I think, 1969 on the Luangwa River while conducting a safari with another professional hunter for four shooting clients; Americans. Fortunately, the result of the incident was a bit more pleasant for me than the simple realization of error as Stevenson-Hamilton. One of my two clients had shot a croc that morning, in fact, so shortly before lunch that I left two skinners with the body to take the belly skin (the only usable part) before it spoiled in the hot sun. Taking the clients back to camp, which was reasonably near, we were having a beer in the dining *kaia* or wall-less hut on a point overlooking the river when the other pro came in with his hunters, having shot a nice pair of bull buffalo. Over lunch, the conversation turned to crocs and how difficult it was for a hunter to correctly judge the size of one before deciding to collect him. No sooner had the subject been broached, than there appeared a pair of the knobs on the surface of the river that mark a croc's eyes as he lies in the current, about 220 yards upstream, near a small bar on our side of the river. As if ordered up, he obligingly clambered up on the mud, turned conveniently broadside and went to sleep with his mouth gaping. I hadn't been paying all that much attention since I'd seen this chap before; the other pro, being a visitor to this particular camp, was therefore unfamiliar with him. He was a lot bigger than what we used to call a "watchstrap" size, but certainly not worth shooting as a trophy by any stretch of the imagination. So, when I heard the other professional remark that yon saurian was about nine feet long, I, without thinking, embarrassed him by commenting that the croc was nowhere that big. It was, of course, a bloody stupid thing to say which tended to belittle him in front of his clients and force him to respond. I have always regretted that day not practicing the old axiom: *PLACE BRAIN IN GEAR BEFORE ENGAGING MOUTH.* But, we ended up as fast friends so it must have not been taken too seriously. In any case, a discussion as to how big that croc really was got underway before you could call for a fresh beer. Binoculars popped out of bush jacket pockets and off gun rack pegs. To save a lot of discussion, all four clients came in with estimates between eight feet six and ten feet even, while the pro dropped his guesstimate to eight feet neat after a hard stare through his glasses. Of course, it wasn't long before somebody suggested that the only way to prove the point was to shoot the croc, to which one of the other professional's men agreed as he wanted just enought skin for a suitcase and attache case. I should not like to compromise anybody's impression of the high sense of morality of the average, everyday, garden-variety hunter, but it would only be fair to say that quite a reasonable amount of coin of the realm, bearing the likeness of Dr. Kenneth Kaunda of Zambia was suitably escrowed. I stuck with my original estimate of six feet two inches and had my fingers crossed as the other bwana and the client went off up the river bank and presently ushered the snoozing croc into paradise with a well-placed 7 mm Magnum that cured his problems forever. A couple of the camp staff zig-zagged through the shallows out to the bar and dragged the deflated creature back to shore where he was stretched and measured. Five feet, eleven and a quarter inches! That came out to almost $300 worth of croc for me, and even I had overestimated the size. That should give some perspective on how big a really huge one must look!

I was never going to tell this tale on myself, but then I never reckoned I'd live this long. But having done so, and believing that I will never again be a professional hunter in Zambia, I now see no harm since, despite the attempt, the dastardly crime I am about to describe never actually took place, despite my best efforts.

In *DEATH IN THE LONG GRASS,* I wrote of seeing what I considered to be two really big crocs, one of which I have mentioned as the 15-foot plus woman-eater Mason killed. The other was the same approximate length, but with what I guessed to be a yard of tail missing. I find, incidentally, in researching crocs through old books that this oddity of missing tail ends on very large ones is as common as not, although I have no idea why. I wrote of having watched him for a half-hour one day when I had no clients and guessed him at what would have been about 18 feet had he retained the missing chunk of tail. A lot of croc in any case, and that's no crock! I also said that I left him alone, which, in the first instance, I did. Further, I indicated that I never saw him again.

I lied.

It was about a week later that Silent, my faithful old Number One, a bushwise Awiza ex-poacher whom I had sent to check a leopard bait, came trotting back only a short time after he had left. Surprised to see him headed back to camp, where my client and I were having after lunch tea, I listened carefully as he described the "Father of All Crocodiles" which had parked itself on the edge of a long, sandy beach on the concession (our) side of the Luangwa. From his story, this croc was as big around as a baobab tree and probably ate bull hippos whole without the slightest hint of heartburn. Only one it could be, and when I saw him, I knew it was the same. But, there was something of a problem. . . .

My client, who has since gone to his presumed reward, didn't know it, but in my book he was ranked in a dead tie for the worst shot I have ever seen. The other guy was blind. He was extraordinary; his marksmanship so exquisitely awful as to befuddle the law of averages! He could miss the inside of his hat three times running if it were propped over the muzzle of his rifle. Worse, the only thing he could do would be to perform a perfect heart shot on an unseen and thoroughly protected female oribi hidden in the grass ten yards behind a greater kudu which would have made the book. He achieved the distinction of not only missing the brain shot on a rather nice bull elephant from 16 yards—*twice,* but then reloaded and missed the entire elephant twice again as the half-dozing tusker wandered away, probably wondering where all the thunder had come from in the dry season! One thing was definitely sure: unerring Elmer, the hunter who did more to single-handedly preserve wildlife than Walt Disney, was not likely to place a precision bullet into that giant croc's noodle at the range of 150 yards, provided he could crawl that far without having a seizure of something else picturesque.

This being my "come-clean," Have-Your-Soul-Simonized session, I'll tell you that I really wanted that croc. I can justify it by reciting the absolute truth that almost without question that big chameleon had eaten more than one winsome, dusky maiden with possibly a missionary and a couple of birdwatchers thrown in. Crocs that big practically have to eat people because they're so bulky they're clumsy and require food as obliging as the indigenous African who keeps filling his or her water gourd at the same spot no matter who was eaten under the same conditions at the same place last week. When you live there, you tend to regard the croc as Africa's own form of built-in retroactive birth control.

But, that would just be a handy justification. I wanted that croc for the same reason those admirable nitwits swim with great white sharks: He was the most impressive killing machine I had and ever did run across in Africa and I wanted to measure him, admire him, count his teeth, open his stomach. I wanted to pat him and say "Wow!" and mean it. I wanted to own him. I wanted to conquer him. I wanted to hunt him because the very idea of the sure knowledge that he would eat me, given the tiniest modicum of opportunity, despite all that education nobody ever had reason to suspect, all the taxes I had paid, all the nice things I had said during my life to people I didn't like, all the nasty dogs I didn't kick; the whole thing wouldn't matter to him. I, in his eyes, would be nothing but a low-calorie snack!

Let's not get any implied nobility mixed up in this, if you please. I did not want to swim with eighteen-feet plus of Ipana smile; I had no intentions of taking the first chance with him. I did not consider him to be awesomely magnificent; visualizing myself as some kind of Africanized Stan Waterman making intelligent, awed, knowledgeable comments about that snaggle-toothed mother being some kind of wonderfully exciting great, mysterious animal. Simply, he scared the bejeesus out of the deepest, darkest corner of my soul, and for that, I wanted to kill him.

Got it? Good, 'cause that's the truth.

Now, how to do it and, equally important, how to do it myself. Clearly, illegal though it was for me to shoot on the client's license, when I saw where he lay, across 300 yards of sand, it was obvious that the safest place he could be was right there, with that particular gentleman shooting at him. So, I took the bull by the prover-bials and explained to my client that thither lay a croc that neither he nor I would see the like of again. The thermometer when we left camp was 118 degrees of October afternoon and it was a lot hotter here on the glaring white sand. As it was, with the heat mirage and the tiny dimensions of that croc's brain box, even the client had a hunch it would have been out of range for both Sergeant York and Annie Oakley. After long consideration and final approval on the part of my client, careful checking through the personnel I had with me to be certain they were all old hands not inclined to air safari laundry with the paid game department informers who drifted in and out of camp, all signals were a glowing green "go." I checked over my .375 H&H Magnum magazine rifle, in those days a custom Mauser built for me by Continental Arms of New York, a marvelously accurate piece with a detachable 1.5X-8X German Nickel Supra scope sight, and thumbed a couple of 300-grain Winchester Silvertip rounds to replace the non-expanding "solid" bullets I usually used. The beach during the October pre-spring low water stretched long and flat ahead, with hardly a feature to provide cover as I slipped out of the shadows and began crawling toward a slight undulation that ended in a tiny ridge about 150 yards from the sleeping monster where I would have a clear shot.

It'll be a long time before I forget that crawl, and it's already been a dozen years. The sand was so hot it was almost unbearable, the contact with my bare shins and forearms drawing an involuntary hiss of pain as the flesh hit the searing grains. Only by slithering my arms and legs slightly sideways to burrow down an inch or so was it possible to stand it, scraping aside the hottest top layer. Yard by yard, I inched on, sweat sheeting into my eyes, tsetses chewing me like a white meat barbeque. God, but that was a miserable time, probably not more than about 20 minutes but it felt like time on the cross. But, so far so good. The couple of ox-pecker birds hadn't sensed me and spooked the big boy, which was more than half the battle. At last, the little ridge was only a few yards ahead—only about a foot high, but enough to provide defilade for cover. Stopping, I rechecked the rifle, snicked off the safety and pressed the top of the scope with the heel of my hand to be sure the mount was securely seated where it should have been.

I never laid claim to being anything but an utilitarian rifleman and I guess that's at least valid or I wouldn't be writing this, but I felt mighty sure that, if I could get a round off from that little hump, it would be fairly unlikely that I'd blow it. I used to spend a big piece of the month of May in the States hunting woodchucks and crows with the .375 H&H before my first safari of the season just to get my shooting eye and stalking skills sharpened, and there's no better way to do it than with the same equipment you'll be using in Africa. If I could do my part, that rifle would sure do its!

At last, I was at the ridge, easing parallel to it so I could squeeze off my shot prone at the proper angle. As slowly as possible, I sneaked into position, slipping the rifle up and the edge of my profile behind it. There he was! He hadn't moved, the individual teeth in his huge mouth looking as big around as your wrist. With the scope power set at 6X, I settled the crosshairs right behind his grin, just aft of his eye, the hold solid as a bench rest at a touch over 150 yards. With the scope zeroed for 200 yards with the 300-grain load (the "iron" express sights under the swing-off mount were regulated for 50 yards and "hairy" work) I held about an inch and a half low to allow for the trajectory and was surprised when with the proper trigger squeeze, the rifle bellowed, the familiar shove of the recoil dropping the big, scaley beastie out of the scope's view. As I brought it back into position a split second later, I expected to see the pink mist of exploding brain matter hanging in the sunlight as the alloy tip of the Winchester slug did its work.

But, I didn't.

There was an odd, fluttering, flopping sound followed by the rattle of something across the river, clattering through the hardwoods. In the scope lens a weird fog of out-of-focus white mist drifted off ghostlike, baffling me. Worst, the reaction of the croc, a bent bow of tremendous arched muscle shocked right off the ground with the boom of the shot. Unhurt, the croc rocketed the few yards to the water and was gone, a foot-high ripple slowly dying as it fanned out in a quieting vee. He was gone. Forever.

Exasperated and baffled, I lay there in the hot sand catching my breath, hopefully staring through the crosshairs at the water in the wasted hope that the croc might yet surface to see what had happened and I would have another chance. Nope. He didn't get that big that way. Soon, Silent with his brother, Invisible, was standing by me, the old man walking forward to mark a place on the beach with the iron blade of his spear, about ten yards from the spot where I had fired. Getting to my feet, I went over and saw the short groove where the bullet had just clipped a tiny rise in the sand, ploughing on another foot to barely catch the top of some kind of a small mound made by an insect or perhaps a mollusk or other water animal. That close to the muzzle, the raised spot wasn't visible through the scope which was focused much farther out. With the difference in actual elevation between the bore of the barrel and the telescopic sight mounted above it, there just wasn't enough room for the bullet to clear. It struck the sand and ricocheted harmlessly away. Gone was the dragon of my dreams; perhaps of my nightmares. I have wondered how many people ultimately passed over those teeth—if, indeed, any—because I missed that easy shot. Well, perhaps some things are not meant to be known.

Another area I have always enjoyed checking in the writings of some of the more imaginative of the "slowly-I-turned, the hot-breath-upon-my-neck" writers is the incredible list of stuff they have supposedly found in the stomachs of crocs. I recently paid $200 for a copy of the quite rare African classic *KILL: OR BE KILLED,* a 1933 effort of the otherwise reliable Major W. Robert Foran, an amateur hunter who served in Africa as a British Army officer and was a close friend of such luminaries of the chase as C. H. Stigand, speared to death by the Aliab Dinka, and Jim Sutherland, one of the first elephant hunters into the Lado Enclave in Sudan and who was eventually poisoned by the locals. Perhaps such exciting company prompted Foran to describe the contents of the stomach of a 23-foot croc he had blown to Jesus in such terms that no reader could possibly be disappointed:

> "On opening the stomach, we found the following list of things: sticks, stones, a woman's foot, a man's hand, some native beads, an assortment of bangles and anklets, the hoof of a waterbuck, the claws of a cheetah, the shin bone of a reedbuck or other antelope, the shellplates of a large river turtle, the horns of a goat, the foot of a calf, and a variety of other strange objects impossible to identify. There could be no doubt about it being a man-killer. I was glad to have been its executioner."

Now, I've spent a disgustingly lot of time mucking (literally) about the stomachs of dead crocs and I must say I never saw the like of some of the recorded literary collections which we are expected to take seriously. I did have the unpleasant duty of removing several portions of a local Zambian lady of color from a croc, but then I saw her being taken in the first place, and that was far scarier. Some animal bones are common, as are odd bits of sticks and other trivia, such as stones. Aside from that, except for one instance of a croc one of my clients shot and another killed by another professional's client, both of which did have human jewelry inside (one wooden bracelet, one brass armlet) I have seen little to write home about, let alone a book. I suspect that human artifacts are passed after a fairly short time exactly the same way that large bones and other indigestable matter is.

Another common report is that crocs can literally snap off a leg or arm cleanly with a single bite. Not so, as a quick look at a croc's choppers will verify. The teeth are round and pointed, clearly designed for holding and not cutting. True, many limbs have been lost to crocs, but almost without exception, this is when rescuers are holding one part of the man and the croc another. The croc will spin over and over, breaking the bone and twisting the flesh until the limb is torn free. Not a very pleasant sensation, I imagine, but I somehow doubt that those who have survived it would be quick to adopt "Take All of Me" as their theme song.

Crocodile attack methods are designed to put these teeth into action in the manner for which they were constructed. I have seen quite a few animals and two people taken, usually when drinking or swimming. One was a full-grown lioness dragged down as she swam the Luangwa. There was plenty of commotion, but she never reached the surface again.

Despite the myth, crocs do not knock animals from the shore or riverbank into their mouths with their tails. Just think it over and you'll see the structural impossibility of this old tale; no pun intended.

Although crocs do make most of their attacks actually in the water, they will sometimes take game and presumably man at fairly impressive distances inland from the shore. I have related elsewhere the experience of being in a leopard blind late one afternoon with a client, watching a troop of impala come down to drink. As we followed the dainty tableau, a big croc rocketed out of the water's edge and absolutely streaked across the ground, catching a ewe by a hind leg before she got over her astonishment. Seconds later, there was no sign of anything but ripples. We later paced off the distance from the shore to where she was grabbed and found it to be a bit over ten yards. I have never seen an attack over such an extended distance from the water since, but my tracker assured me at the time that it was not especially rare.

I'm just as happy to tell you that I have no personal knowledge of the dreaded salt water crocodile *(Crocodylus porosus)* which ate somewhere near a thousand Japanese troops in a mangrove swamp one night during World War II (and may have eaten Michael Rockefeller off New Guinea); nor am I going to take your time and mine to get into the various South American jacares and caimans, let alone the Asian mugger and Estuarine crocs. These are relatively less numerous than the Nile crocodile and, although both the *porosus* and the mugger have fearsome reputations which we have no reason to believe undeserved, just don't eat as many people as do the Afro crocodiles.

The American alligator is another story and, where any straight thinking, logically minded person would expect to find him in this chapter, tough luck. You'll have to read on a bit as he's a separate case I'd like to discuss elsewhere.

I have never been noted for being very smart, and you will find no legends lurking in any of the bars in Africa concerning my great enthusiasm for bravery. The more I got to know the bush as my career proceeded, the more careful and lucky I became, given that caution is the greater part of luck, anyway. So I relate the following episode as ranking approximately number two in my repertoire of recurring nightmares, a classic example of personal stupidity of the highest order rather than any attempt to impress anybody with the hairiness of my chest.

It was one of those mutually agreed upon feast days, a Sunday, when safaris sharing adjoining hunting areas visit one another for rest, a few hairs of the hyena and some yarn swapping. As usual, a shooting contest was part of the festivities. I was in Mwangwalala Camp on the Luangwa, doing a joint safari with Brian Smith, and we were hosts to Ken Woolfrey and his clients from upriver. This beautiful camp was situated on a high bank over the river. There was a sandbar about 100 yards away, near the far shore. A big, deep pool separated us from the game reserve on the other side, a pool we well knew to be literally crawling with crocs.

When the question arose as to where we would place the tins we planned to use as targets, it was apparent that the thick surrounding bush did not afford enough room for a range. So somebody decided that the sandbar would be the ideal shooting gallery. Now, there was no way you were going to find anybody on our combined staffs dumb enough to swim over there and back if you had been offered gold sovereigns. Somehow, I got elected. I know I didn't volunteer, but all of a sudden, I found myself stripped down to my bush shorts. Carrying a plastic bag of vegetable tins, I stepped reluctantly into the water after Brian had fired four or five .458 rounds into the pool to shoo away any unwelcome residents. I gripped the bag in my teeth and struck out for the bar, every second expecting something very toothy to grab me around the middle. Nothing happened. I waded up onto the bar, set up the tins and swam back with the empty bag tucked into my waistband. Drying off, I was advised that it would only be right if I took the first shot. I cranked up the old .375 H&H, locked in the ready sling and looked over the sights. And practically fainted. Ten yards from where I had crawled out of the river below camp were a pair of 13-foot crocs, looking very hungry and frustrated. They had to have been there all the time! Why they hadn't nailed me, I have never been able to guess. I can promise you, though, that was my last swim in the Luangwa.

Personally, I hate crocs. The reason is that I fear them. There could be hardly an end more horrible than feeling that death-grip of terrible teeth, knowing that there was nothing you could do to save yourself. Still, crocs are as much a part of things as the mambas, puff adders and malaria. They're a part of Africa, and that's good enough for me.

CHAPTER FOUR

LEOPARDS

Among the man-eating cats of the world, the leopard is absolutely unique. The smallest, yet possibly the most awesomely powerful pound-for-pound of the individual animals that consistently kill and eat man, virtually every hunter with personal experience, professional and amateur alike, agrees that a man-eating leopard is the most difficult and dangerous of all the cats to hunt.

There's a good reason for this general, if rather uncheery agreement, too. The pro game ranger or the unpaid nitwit who takes after a man-eating lion or tiger generally either succeeds in killing him or he doesn't. For sure, some hunters have the tables well and properly turned on them through ineptitude or even bad luck, but in the life and death contest between man and man-eater, in no case is the human hunter more likely to end up as the killer's deliberate next meal as with the leopard.

It may be that the lesser amount of contemporary big game hunting literature compared with that of 50 years ago (despite its recent resurgence) has led to the general impression that man-eating activity is in a steep decline. One needed only to be with me in the bush of Botswana's Ngamiland a few years back when I was stalked by the man-eater who killed and partially ate the son of my skinner (of which I've written elsewhere) to realize that there are still, despite what you may have heard, no barnacles growing on the man-eating leopard as an active member of the species. Time after time, those who have either hunted a man-eating leopard or even researched the career of one remark—more than casually—on the uncanny way the leopard has of turning things about.

Colonel Jim Corbett, that intrepid hunter of India's man-eating cats, speaks several times not only of his own fear and nervous exhaustion becoming so great that he was forced to quit his campaigns against both of the two great man-eating leopards he took on—with over 525 human kills between them—but also of the other hunters he knew who had been dragged, gagging with terror from the blackness of their own *machans,* tree platforms on which they waited for a shot over the tooth-torn remains of the leopard's last human victim, themselves unwittingly the hunted, and then the prey. Jack Denton Scott, a fine outdoor writer, in his book *SPEAKING WILDLY* mentions an Indian leopard who had killed some 100 women and children while Scott was in India in 1958. The cat had also killed and eaten four professional hunters sent after him by ambushing them in their tree blinds. In retrospect, what the really experienced man-eating leopard learns to do is to actually use bait for the *hunter!*

Think about it. . . . A leopard learns if he leaves part of the body of his last kill somewhere that it will be found, he soon discovers that, like clockwork, somebody will be sitting up over the body waiting to kill the leopard upon its return. The cat follows this idea one step farther, though. Instead of returning to feed upon the body, he may ignore it in favor of the hunter he knows will be nearby. With the leopard's perfect night vision, it's as easy as sniping with an infrared scope for the cat, who eases up with the greatest hunter's stealth in the art of death and takes another victim: the hunter.

A mere glimpse is usually sufficient to establish the physical qualifications of either a lion or a tiger to eat just about anybody they want to. They're both big, strong and mean. But, to really appreciate the combination of skill, boldness and extraordinary technique of the leopard as a man-eater, a little closer look at both the physical and mental makeup of the animal is worthwhile.

Of all the dangerous, or, come to think of it, even nondangerous big game animals of the world today, none is more widely distributed geographically (except possibly the wolf) or certainly none in better harmony with a wider variety of habitats and unrelated species, including man, than the leopard. The British used to call him a "panther," especially in Asia, but, despite a hundred arguments involving as many imagined differences in physical characteristics as enumerated by paw prints, rosette conformations and skull measurements, he is the same critter no matter where he's found, whether from southern Africa to the frozen wastes of Russia and Manchuria, to the golden beaches of Malaya, Sumatra and Java or the searing Sinai Desert of Israel, so tantalizingly close to Europe. True, there are at least 15 separate races recognized, but he still is none other than *Panthera pardis.* He is equally at home in the below sea level depressions of Africa as in the freezing heights of 16,000 foot mountains. Remember, after all, Hemingway's remark as to the frozen body of the leopard at 18,000 feet on Mt. Kilimanjaro and the question of what it was doing there in the first place. Looking for something to eat, I suspect, if he was true to his clan; quite possibly trying to nab a mountain climber.

Above any single facet of the leopard's makeup, it is his adaptability that has proved most effective to his success as an achiever in a world that so dearly wants to stamp him "endangered" or "nearly extinct" despite the fact that he's in about as good shape as any predator in business today. As this is being written, the leopard is about to be readmitted as a game species which may be imported into the United States after years of banning. Why? Easy. There are *too many* leopards, especially in Africa where they're causing much livestock damage. Since they received international protection from foreign import, stock owners are forced to poison them because they no longer generate income from hunters sufficient to justify their existence, although local hunting is completely legal. But, what visiting sportsman will pay what may aggregate nearly U.S. $1,000 in licenses and fees when his trophy cannot be imported to his home country?

Most people who have never had the privilege of spending any time in the range of the leopard under circumstances that would permit some observation of the animal under natural conditions are amazed to find what a small, yet what a savagely powerful animal he is. Of a certainty (which is a phrase I always hoped might give my writing some dignity if I ever found a place to use it) 99.99 percent of the people who *do* live in leopard country, even in areas of heavy population densities of the animal, may spend their entire lives never catching the slickest, sun-mottled glimpse of *Tendwa* or *Chita-bagh* as he's generally pegged in India, or *Chui* or *Ingwe*, *Nyalubwe* or any of the other dozens of African names he sports. He may pick off your loving wife; the daughter whose bride-price will keep you solvent during your declining years until your visit to Tomorrowland or your son and heir any dark night they get careless and he gets lucky. You will most likely never have seen him although he's lived his whole life within a mile or two of your hut. He's almost part of the village family, the answer to where the dog that "ran off" last month went; what happened to so-and-so's missing chickens. He's sorted through your garbage a hundred happy times. The temptation of that plump, helpless child or the unsuspecting woman in the dusk was just too much. Like the casual killer he is, he struck. Silently, neatly, just a short hiss of shocked breath and the crunch of fangs on bone, and that was that. He may never kill another human or he may take a shot at the record of the Panar leopard at better than 400. Either way, he's just doing what comes naturally.

It would take about four normal, male leopards to equal the weight of an adult tiger and three to counterbalance a good-sized lion. At an average of well under seven feet, including a couple of which is tail, the leopard is almost insignificant compared with the really big man-eaters such as crocs, bears, and the many other species we deal with in this book. Not only for his size, but overall, there is probably not another carnivorous animal on the face of the earth more intrinsically deadly if he picks up bad table manners than the leopard. Compared with the other land-dwelling man-eaters, he's as hard to hunt and kill as to catch ground fog in a narrow-necked bottle. He may never return to a kill once he's fed upon it; on the other hand he may, but as hunter—not hunted. Even though he does return, don't think that like the lion or tiger this will necessarily allow a shot. He is *so* cautious and so goddam smart that, for just one example, the Rudraprayag leopard, which Jim Corbett hunted in the Himalayan foothills of Kumaon, came as close as two feet of the armed and ready hunter on at least four occasions and was not shot because of one intervening object or another.

"Man-eater" is almost an inaccurate term as applied to the leopard because such a large percentage of his normal human victims are smaller women and children. This is by no means a hard and fast rule, as some leopards have carried a full-grown man weighing over 150 pounds more than four miles after killing him, but, because of the obviously lesser fuss of killing smaller people, the bulk of those taken by leopards are in this category.

I do not share the opinion of some hunters and writers of man-eating leopards such as Corbett, who maintain that any human kill that took place at night was the responsibility of a leopard and in daylight that of a tiger. In Corbett's personal experience in Asia (he later moved to Kenya Colony in East Africa but had no chance to hunt man-eating lions who will grab a person any time they feel the urge, although operate mostly at night) this may have been the fact; he specifically says it was. My objection is that of his explanation of this phenomenon.

". . . Both animals are seminocturnal forest-dwellers, have much the same habits, employ similar methods of killing, and both are capable of carrying their human victims for long distances. It would be natural, therefore, to expect them to hunt at the same hours; and that they do not do so is due to the difference in courage of the two animals. When a tiger becomes a man-eater it loses all fear of human beings and, as human beings move about more freely in the day than they do at night, it is able to secure its victims during daylight hours and there is no necessity for it to visit their habitations at night. A leopard, on the other hand, even after it has killed scores of human beings never loses its fear of man; and, as it is unwilling to face up to human beings in daylight, it secures its victims when they are moving about at night, or by breaking into their houses at night. Owing to these characteristics of the two animals, namely, that one loses its fear of human beings and kills in the daylight, while the other retains its fear and kills in the dark, man-eating tigers are easier to shoot than man-eating leopards."

Speaking personally, I have all the respect in the world for the late Col. Jim Corbett, both as a gentleman, a talented hunter of man-eaters and a bushman. But, I think he's gone too far here, and that would include many facets of his opinions.

First, whether it was Corbett's experience that *all* night kills were by leopards and *all* daylight kills by tigers, we will have to take his word for, although he himself says that there are great similarities in the killing methods and a mistake or two might have been made, even by the *Maestro*. If he says that it was his experience, though, it's okay by me. What I find uncharacteristic of Corbett is his application of what are, after all, strictly human values and value judgments to what are purely animal behavioral patterns. It may be that two very bad man-eating leopards Corbett hunted (the Rudraprayag and Panar leopards) came close to killing him or forcing him into a mental breakdown by the time he wrote his books that he had developed a real hatred of leopards sufficient to insult them in print with words such as lacking "courage" and "retains its fear." It's almost as if there were some secondary moral value to the animal's actions other than to the *modus operandi* of the cat himself, which Corbett admits of the two, the leopard is the more effective. Of what worth to a hunting, man-eating leopard is "courage" in a human sense if it would only make him a better target in daylight and reduce the effectiveness of his foraging by being seen? He's abroad to eat people, after all, not to show off his nobler qualities!

And, if indeed, "courage" or the lack of "fear" is an implied virtue in man's oldest enemies, the animals that would eat him, what is there really for us to cheer about because the tiger makes his kills in broad daylight, making new widows and orphans in the clear light of day instead of the murk of night?

Not many—if any, indeed—shared Corbett's implied loathing of the leopard, calling it "a scavenger" while the tiger was to him "a large-hearted gentleman with boundless courage. . . ."

The fifth generation Anglo-Greek game ranger of Tanganyika who became as well-known for killing more than 25 man-eating lions and leopards as for his expertise with poisonous snakes, C.J.P. Ionedes, regarded the leopard as "a bolder animal than the lion—the leopard will, for instance, go into an occupied house and take a dog . . . A good male specimen averages 120 pounds against the lion's average 400 pounds. Yet, in my experience no African animal is more difficult or dangerous to deal with than a leopard, or a more formidable proposition when it starts preying on humans."

Not just Ionedes had a real respect for the leopard. The experienced hunter and writer Major W. Robert Foran thought, "In point of pluck, cunning or ferocity, the leopard is the peer of either tiger or lion. . . . Leopards have more cunning and courage than any other animal I have encountered in my hunting days." Again, some humanization, but certainly from a different view of reference!

There being no point in flogging this matter to death because we will shortly see why Corbett was not overly fond of leopards, I would just like to mention that there is a vast fund of difference in others' experiences as far as Corbett's day/night conclusion and the inferences he draws from them are concerned. I, myself, know of four African children and one woman killed in the daytime by leopards, one child nearly at noon while the others were either in the early morning or late afternoon when most carnivors are hunting. Of course, the majority of kills by leopards are at night. Theodore Roosevelt in *AFRICAN GAME TRAILS* makes a point clearly about leopards killing people in daylight, however, quoting District Commissioner Piggott of Neri, Kenya Colony: "He told us that at the same time a man-eating leopard made its appearance, and killed seven children. It did not attack at night, but in the daytime, its victims being the little boys who were watching the flocks of goats; sometimes it took a boy and sometimes a goat. Two old men killed it with spears on the occasion of its taking the last victim."

The above report by Commissioner Piggott leads handily into a bit of a discussion into the reasons for the leopard taking up people-eating and the huge differences between the patterns of hunting and feeding between the big cats compared to the leopard. We've already investigated the rest of the general classifications of man-eating animals to the point that, if you're not sick of it, I sure am. But, the leopard is the only one of the higher life forms that seems to go right along, at least to one degree or another, treating man as if he were part of the normal balanced diet. You know, the recommended daily allowance of Riboflavin, iron, etc. A very, very few leopards will *only* eat man, passing up every other meat source available, although this phenomenon is much more common with lions and tigers.

To be fair, Jim Corbett put the cause of man-eating among leopards as a direct result of scavenging on bodies which gets out of hand when the unrestricted slaughter of the normal population of prey animals deprives leopards of their usual fare and they are driven to man-eating, or an epidemic achieves the same result. He points out that one of the killer leopards of Kumaon *did* follow the great influenza epidemic of 1918.

Jim Corbett, however, was of the old school of presupposition that any animal who would eat man in the first place must be "depraved" or in some other way subject to some kind of trauma responsible for such behavior, injury precluding the pursuit of more standard prey the favorite reason. In the matter of tigers, all of Corbett's man-eaters had been either injured by a man or were suffering from porcupine quill damage that had so badly crippled and disabled them that the hunting of normal prey was impossible. To give Corbett his due, and remember that his sample is not nearly large enough to be valid as a test or control group in a statistical experiment, a significantly higher percentage of man-eating tigers have been incapacitated to one degree or another than either lions or leopards. His conclusion of tigers, however, by no means assures that his assessment of leopards was correct.

As a matter of fact, a thorough study of 78 man-eating leopards at death in Peter Turnbull-Kemp's classic work *LEOPARD* (Howard Timmins, Capetown, 1967) shows that 90 percent of the leopards operated in areas where game or domestic animals were both common or present and 73 of the 78 were in at least "good" condition (presumably meaning uninjured) which is a walloping 93.6 percent. Right. Now you tell me that leopards don't eat people as a regular part of their diet. Oh, incidentally, out of another grouping of 152 known leopard man-eaters, only nine (or 6 percent) were females.

Having spent a portion of my life in the hunting of man-eaters, especially lions and leopards as far as cats were concerned, I always had a fairly reasonable justification for wondering just what the odds were of escaping from the clutches of a man-eating lion or leopard once a person was actually attacked.

In a word, slim. In two words, very slim.

Out of 125 certified kills, the Rudraprayag leopard had two survivors, one man and one woman who managed to tear loose and escape. The Panar leopard "lost" one of his official 400 victims but she died shortly thereafter of septacemia. The Mirso leopard, who killed over 100 people in the Golis Mountains of Africa is reported as completing all innings without error. In 1937, the Chambisi man-eater impaled itself by accident leaping upon his 68th victim, who was carrying a spear which fatally pierced the cat. In my old stamping grounds of the Luangwa Valley of Zambia, one year later than the Chambisi chap, a leopard went straight, simply biting the throats out of his victims without feeding upon them. In roughly 1960, Ionedes reports the same phenomenon taking place on Tanzania's Ruvuma River when 23 souls went a-winging from Masaguru Village before Brian Nicholson, Ionedes' protege, shot it feeding on the body of a dead elephant. It had not eaten a bit of its human kills either. So, I would suggest, friend, that if a man-eating leopard rings your bell, you might as well buy a subscription.

The human-hunting technique of the leopard is, despite the accusations of "cowardice" by Corbett, really almost exactly that of its normal pursuit of other game. Of any book one reads about the killing methods of big predators, there will always be a certain amount of discrepancy between authors based upon individual experiences; an excellent example being descriptions of how an African lion kills a buffalo. Some very experienced hunters and naturalists claim that the neck vertebrae is dislocated or broken by making the buff fall in such a way that his own weight accomplishes this. Others note that the anchoring bite is in the back of the neck while others swear that their eye-witness sightings are clearly in the front of the throat, in a throttling manner. Personally, I have seen all three results, so take your choice. In the case of the method a leopard uses to kill a person, I have seen and inspected wounds in the back of the neck, crushing the third or fourth vertebrae with the canine teeth as well as a clean frontal throat grip which I imagine scrunched the larynx and rather quickly caused death by suffocation. But the most common death-bite, which should not be confused with the extensive damage so gleefully dished out by a *wounded* leopard to a following hunter, which would be the anchoring of the teeth in the face, neck or shoulder while the hind legs flash like a pair of super-charged Roto-rooters in overdrive to claw out the stomach and lower intestines, would be that used most commonly on baboons. It's a savage paw-cuff followed by a lightning lunge at the base of the skull which drives in the four canine teeth as neatly as a carpenter would hammer spikes.

That particular bite happens to remind me of a story which just might fit nicely here, since we're discussing the hunting stealth of the leopard as well as the near-supernatural skill of his technique. The incident took place in August or September of 1975 in northwestern Rhodesia, in an area bordering the Wankie National Park about 40 miles south of the Victoria Falls, at a concession known to the Ama-Ndebele and BaTonka people as *Mongu,* for a small stream that ran through the place. I was so fortunate to have as my "client" (I always feel slightly odd referring to one of the world's greatest hunters and shots as a "client") the famous writer, aficionado of most things worth being eaten, drunken of, tracked or generally ad-mired, the full-time gentleman, Colonel Charles Askins. Aside from some off-pre-dilection for an overwhelming obsession to fire 8 mm bullets faster than either God, Congress or Roy Weatherby meant them to go, Charlie Askins is just the chap you want to be able to whistle up in a hurry in case of a latent hangnail, charging buffalo or other lurking tragedy of the African *bundu.*

Charlie and I were out one afternoon in the rocky hills some hundreds of feet above a *vlei* or dry grass field in what ultimately turned into a successful attempt to place a fatal hole in a record book sable antelope. The sun was starting to drop to the point where the shadows were lovely and cool as we picked our way along the well-grassed and bush-studded basalt hill we were hunting, aware of a troop of baboons feeding a few hundred yards down-slope from us. As if at a given signal, the entire troop exploded into vocal hysteria, which we both knew probably meant one thing: leopard. And, by the way the screams stayed static as opposed to the way they would move if *Ingwe* had been spotted at a distance, it seemed likely there had been a kill made by the cat. Although Charlie had no leopard license that trip, we both started running downhill to get a better look at the proceedings, somewhat against my best judgment as I have had enough years in Africa to naturally shy away from close encounters of *any* kind that had anything to do with leopards.

As we got farther down the slope, we could see that the baboon troop was a very large one, perhaps 60 or 70 strong. What bothered me was that nearly all of them seemed to be in a couple of taller trees on the slope, alternately staring from us to a spot on the hill about thirty yards below them. It didn't take Sheena, Queen of the Jungle to figure out that they were watching a leopard who was still on his kill, and were too scared of both us and the cat to decide whether to run from the tree or stay put. I studied the whole area closely with binoculars but could not see a whisker of either the leopard—if, indeed that was what it was—or the kill.

With Charlie and I covering each other as best we could, feeling a little silly because the bush was so light, we started to ease toward the spot the baboons were shrieking at. Near the base of the hill where it abutted a dirt path, there were, I noticed, a couple of shallow gullies, sluiced out by the previous spring's rains. This seemed to be the place that held the baboons' interest.

The ground, except for an occasional tuft of low grass or dead branch, seemed as incapable of hiding a shrew as a leopard. Sure the cat had gone somehow across the 30 yards or so to more dense cover before we had come up, I went straight toward the gullies, not deeper than a foot and perhaps eight inches wide. When I was within *six feet* of the nearest, I slowly turned my head to check on Charlie. As I did so, there was the tiniest slithering rattle of a pebble on the dry earth which my darting eyes flashed back quickly enough to spot. There, in what I can only call an impression of motion, was the dappled disappearance of a big—very big—leopard. He was long gone before I could have even dreamed of lifting my .375 rifle, let alone getting off a shot. A closer look showed a tiny gleam of reflected blood from the fur of a half-grown female baboon which the leopard had been shrouding in the ditch, as close to me as I was tall! Charlie and I glanced at each other without comment.

In retrospect, that was nearly as close as I have come to being hashed by a leopard, and that includes some other most impressive encounters. There is no question in my mind that, had I shown signs of having seen him he would have felt cornered and charged. From six feet I wouldn't have had the chance of a mouse with a mamba of avoiding a spontaneous subdivision. Another step, and that would have done it. What's the point? Well, that leopard was no man-eater and was trying to put some miles between us, not sneak up as a people-eater would have. He was practically wearing a sign the way the baboons were staring and screaming at him, but I still couldn't see him from six feet off! What chance would I have had if he'd been after me rather than just anxious to get away?

The bites, incidentally, through the skull of the baboon were identical to those which killed Xleo, the son of my Botswana skinner, July, which I reported in

DEATH IN THE LONG GRASS with the exception that these were a bit higher, through the skull itself, in that area I believe is called the *medulla oblongatta*. In any case, the result was certainly the same, instant death.

It seems only reasonable, especially since we broached the subject earlier, to examine the case of the animal that may have been the single recorded highest scorer of any man-eater in history, the Panar leopard. Now, many devotees to the subject of man-eaters may be quick to protest that the "official" score of the Panar leopard, which Jim Corbett killed in northern India, was only 400, 36 less than the Champawat tigress (which Corbett also shot). But the point remains that really accurate scores were never kept on hill people who simply turned up missing or were discovered dead after too great a degree of putrefaction had set in to identify the precise cause of death, whether a man-eater or a fall, followed by attendance upon the body of scavengers.

By way of example, I offer the three best known of Corbett's man-eaters and their "official" scores:

The Rudraprayag leopard—125 kills
The Panar leopard   —400 kills
The Champawat tigress   —436 kills

These numbers are just too "pat" to be believable on the high side. Corbett tells us himself that, during the period he was hunting the Rudraprayag leopard, there were many human kills he knew of which were never subsequently recorded for whatever bureaucratic reason one chooses to believe.

Undoubtedly, and Corbett confirms this, the same was the truth with the Panar leopard. I am suspicious of any animal who kills precisely 400 souls, stopping on such a conveniently even number. Did the man-eater send the government memos reminding them of any victims they had missed? The Champawat tigress is even more suspect, although on the high or low side is anybody's guess. I believe the estimate was high, personally, since the cat is recorded as having been chased into India from Nepal after it had killed, based upon whose figures we have no idea— also exactly 400 persons. These tales do not usually lose victims in their retelling. For my money, on the other hand, the Panar leopard was probably high scorer overall.

Although Jim Corbett, an army officer associated with the Indian Railways, had been born in the Himalayan foothills of Kumaon in 1875, he was 32 years old before coming across his first man-eater of any description, the Champawat tigress in 1907. While hunting her, he first heard of the Panar leopard terrorizing the Almora district but did not start his attempt to shut down her operation until April of 1910. In fact, to substantiate the inaccuracy of the numbers of victims of this animal, Corbett himself says that no government bulletins were issued or records kept between April and September of this year although certainly a good number of people were eaten.

Jim Corbett's first try for the leopard in April was to be one of the more frightening episodes of a career not known for its dullness. After having hunted down and killed a man-eating tiger known as the Muktesar man-eater in the area of that same name, Corbett and his hill men walked for several days into the Almora district, setting up a simple headquarters at a place called Dol, where there was a *dak* bungalo or government rest house for persons on official business. The next morning, picking his way along a path connecting a series of small villages, Corbett cut the track of a big, male leopard and, a short distance farther on, saw a lonely stone house on a piece of scrub farmland carved from the jungle. As he approached the house, a young man, incoherent with panic, burst wild-eyed out of the door, down the stairs and across the small courtyard to Corbett.

The previous night, while the young farmer and his wife were sleeping in—if you can believe it—a room with the outside door left open, the leopard had sneaked in and grabbed the woman by the throat, starting to drag her outside until her strangled attempt at a screech awoke her not over-bright husband. With commendable bravery, however, he grabbed her arm with one hand and braced himself against the door frame with the other, with a lurch, ripping his wife literally free from the jaws of death. Somehow, he was able to slam and jam the door, although the leopard, infuriated at having lost his meal, began trying to tear it down. The rest of the night was spent in the sheerest terror, listening alternately to the strangling gurgles of his wife struggling for life breath with her mutilated throat and the growls and rasping claw scratches of the man-eater trying to break in.

By dawn, the husband was about at the end of his string, his wife unconscious and, by the smell her wounds already turned septic in the damp, hot air. The young man was panicked, torn between trying to run for help to the next house, more than a mile away through the thick jungle, and staying where he was, trying to somehow save the dying woman. Since dawn the leopard sounds had stopped, but he could not tell if the big cat was simply waiting silently outside for him to open the door a crack or had left. When Corbett *sahib* happened to track the pug-marks past the house, it was just in time to keep the husband from stripping what mental gears he had left.

Corbett knew that the nearest medical aid was back at Lamgara, some 25 miles away, far too distant to give the woman a chance to survive the trip. In fact, when he was able to inspect the terrible fang holes in her throat and saw that they were already gangrenous and that her breast was ripped by the claws and badly infected too, he knew for a certainty that there was no hope for her. Toughened as he was from a lifetime of the rank realities of rural Indian life early in this century, Corbett was nonetheless too queasy to stay in the close, steaming room that smelled like a combination between a charnel house and a septic tank because of the rotting flesh and the lack of the most basic sanitation. He sat up all night, back against an outside wall, waiting for the leopard with his rifle, listening to the mangled woman choking with the swelling, rotting wounds. Afraid that the *sahib* would surely be taken by the man-eater, the husband asked what he should do in that sad event.

"Close the door," suggested Corbett dryly, "and wait for morning."

At first light, the Englishman was off to send back help, academic though it would be. The man-eater must have moved on in search of easier pickings as there had been no sign of him during the night. When medical aid did at last arrive, the girl was as dead as easy credit and, with the man-eater's trail cold, there was nothing Corbett could do but wait for another human to be killed to take up the hunt again. As it turned out, with professional complications, that would not be until five months had passed, the following September.

On the 10th of that month, other affairs squared away, Jim Corbett left his home at Naini Tal at four a.m. and walked with his men the 28 miles through the rain to the Almora district border. With the intention of following the same general route as his hunt in April he kept on for another four days, inquiring at each *patti* or small group of villages for any news of the leopard, hearing that there had been a human kill at a place called Sanouli, on the far side of the Panar River which, because of the heavy rains, was in flood. There was nothing to do but wait for the waters to subside.

Not worried that the room he chose in a long row of typical two-storied buildings had no door, as the leopard would not cross the flooded river, Corbett had his groundsheet spread and his bed made up. Eating his dinner seated on the cot in the small, oddly bare room, he dozed off, tired from the march and slept soundly, to awake to a terrible shock. As the first early rays of sun glowed through his eyelids, he opened them to notice a figure seated right at the foot of the bed. Corbett's skin crawled when he saw that the figure's face was disintegrated with the final stages of leprosy! Explaining to Corbett that this was his room and that he had been away for two days, he had not wished to awaken the *sahib* so just sat at the foot of the bed waiting so as not to disturb him. The local people, as Corbett put it, were rather fatalistic about the dread disease, but he spent the rest of the morning scrubbing his body and equipment practically medium rare with carbolic soap.

After a very tricky crossing of the river, in which one of Corbett's men nearly drowned but was saved by another (whose new coat he had strapped to his back) the obstacle was negotiated and Sanouli reached at about noon the next day.

A very small and primitive mountain village, Sanouli was built on a hillside over- looking cultivated fields, across the valley from which was a 20-acre stand of thick brushwood which stood islandlike between the open fields and natural grasslands. Somewhere in this tangle of snarled vegetable crud, all present agreed, was the man-eating leopard who, in addition to the last kill, had also taken four other people between the last half of March and the first two weeks of April. Now, the cat was back and, if he behaved as before, the latest body would be about 500 yards into the patch of brushwood where the earlier victims had been eaten. After a walk around the perimeter of the wooded area without spotting any exiting pug-marks, Corbett bought two goats and staked them out that night, sitting over one while the other was killed by a leopard who dragged it back into the brushwood patch.

Having in my errant youth in Africa had occasion to do so also, I must agree with Corbett's observation that "stalking a leopard . . . on its kill is one of the most interesting forms of sport I know of. . . ." However, the cover was too dry and the colonel had no luck (neither, at least, did the leopard) despite being advised by the excited calls of such birds as babblers and drongos of the presence of the cat in the thicket. He considered organizing a drive to beat the killer out of cover, but knew from experience that this would probably only cost the beaters an impressive collection of casualties. To top off events, Corbett went down with recurrent malaria which put him quite off his head for a day and a night. When he recovered, he learned that the surviving goat had been staked out by his men but had not yet been found by the man-eater. Oh, brother, Corbett probably thought viewing the only tree of any description near the brushwood, a disreputable and mostly rotten oak growing from a six-foot bank between terraced fields at an angle shallow enough that the hunter was able to walk up it in his rubber soled shoes. What a swell place to spend the night on the only branch, a rotten one at that, just a foot wide on the underside of the tree and but 15 feet from the ground, waiting in the dark for a leopard that's eaten over 400 people! Tying a white rag around the muzzles of his shotgun for a night sight, Corbett decided that some people have all the fun!

From this you just may have gotten the idea that Corbett is a tiny bit rash. But, not so. To protect himself, he directs his men to cut bundles of tough blackthorn shoots and saplings to tie tightly to the leaning tree so the leopard can't walk up the trunk the way the man did. Having arranged his vegetable barbed wire, Corbett takes the further timely precaution to turn up his coat collar so the leopard, should he manage to get past the thorns, will have presumed difficulty driving his fangs through the back of the man's neck! Since the blackthorn shoots are very long and there being nothing else to hang on to on the branch protruding from the underside of the leaning tree, Corbett, presumably after trimming some of the tips of the shoots of thorns, tucks them under his arms and, pinned against his sides for balance, settles down for an uncomfortable wait for the leopard who has probably been watching him for the past half hour from cover anyway.

There are places Colonel Edward James Corbett would rather be. . . .

As the last blush of daylight fades from the rocky slopes of the lonely foothills gripping the tiny village of Sanouli, Corbett gives an involuntary shudder of horror, as he feels the blackthorn shoots twitch under his arms. Unable to shift far enough around to bring his shotgun into use without knocking himself off his branch, he realizes that the leopard is no more than 15 feet away lurching, pulling and twisting at the saplings to shake the man free. With an especially savage lurch punctuated by a rumbling growl, the Panar leopard stares upward with eyes dilated wide to catch the dying light. It's going to be a long, long evening, Jim Corbett thinks.

Another sharp growl washes over Corbett from a few feet below and behind him as the leopard suddenly releases the shoots he has been pulling upon, nearly flipping the hunter out of the tree with the sudden slack of tension. The blackness is almost total, the only things Corbett can see are the faint whitish blur of the goat thirty yards away and the pale glow of the rag wrapped around the shotgun's muzzle. But, at least, as long as the man-eater is growling, Corbett knows where he is. What if he turns silent and decides to leap upwards from directly beneath the branch? If he's even brushed, Jim Corbett knows he will be knocked from his rotten perch to earth. Then, for some inexplicable reason, the Panar leopard stops trying to pull him out of the tree and rushes the goat!

A gentle man, Corbett has tied the bleating animal in such a way as to give him time to kill the leopard before it reaches the bait. But, not in the complete darkness and at a time before shooting lights were known in India. The goat must die; it does. As it stops struggling in its death throes in the leopard's jaws, Corbett matches the white of the cloth mark against the slight gleam of the dead goat and touches off the right barrel of buckshot. Through the flash of the muzzle blast he has an impression of the leopard giving a grunt, turning over and disappearing over a ledge into a field below. In complete blackness, the night settles back into silence.

After a quarter-hour, Jim Corbett thinks it safe to answer the calls of his men, telling them to light the long, thin strips of resin-rich pine that are the traditional torches of the Kumaon hills. So hard has the leopard tugged at the bases of the bundles of blackthorn that the ropes holding them must be cut before Corbett can climb down from the tree, which he does, reluctantly agreeing to look with the men into the field below and past the dead goat.

With exaggerated stealth, the 20 men approach the edge of the raised terrace, Corbett having been assured faithfully that if the cat isn't dead and charges, the hill men will not run away and leave the *sahib* in the dark. No, of course they won't. . . .

As the last band of shadow evaporates in the quavering light of the pine splinters, there comes a furious series of ferocious grunts as the wounded man-eater charges from the field and up the bank, straight at the knot of men. Instantly, panic strikes the torchbearers as they recoil in terror, some flattened by collision with others, their flaming brands falling to the ground to sputter an eerie light. In the lead by five paces, Corbett swings up the shotgun trying to center the racing, golden blur of charging leopard. At point-blank range, a swarm of SG pellets churn into the cat's chest, smashing him into a limp rag of flesh. Jim Corbett has won. Not by a helluva lot, but it's enough. At least he is not angry at his men. Remembering his own deadly fear in the tree just a few minutes ago, he's man enough to observe that, had he been one of the torchbearers, he would have passed the rest of the crowd before very many steps had been taken.

What does it matter, anyway? The Panar leopard is dead.

CHAPTER FIVE

BEARS, WOLVES AND HYENAS

O n a geographical basis, it seems to me a fellow would be hard put to find a more widely distributed form of terrestrial omnivore than bears in general. All sharing remarkable similar characteristics beyond such cosmetic considerations as color and size, bears are found in different flavors just about everywhere but in Antarctica, and if the polar bear ever was able to get past the equator, there's little question that the bottom of the earth would not be bearless.

This not being a zoological reference book, I can't believe that the total readership interest in the South American spectacled bear or the Asian sun or black bears would be worth the calories expended to include them, although Lord knows I have the spare calories. So, it is my dearest hope that you don't feel shortchanged by only being violently dissertated to on matters pertinent to grizzlies, browns, polars, blacks and sloth bears, any one of which, believe me, is better encountered in these pages than in its natural habitat unless you have an emerging death wish.

With the possible exception of the great cats, there probably has been more speculation, both correct and otherwise, written about bears than any other animal group. With good reason, too: they're big, scary looking, and they bite. A recently processed pile of still-steaming blueberries encountered in heavy cover has an astonishingly stimulating effect upon the sense of foreboding of just about any hunter, fisherman, hiker or backpacker you can think of. There is probably a case to be made that this natural apprehension is a residual reaction to the days of yore when we started to solve the housing shortage by some very rude treatment of Pleistocene cave bears, whose rocky homes we undoubtedly usurped. The most recent thinking on the matter, incidentally, indicates that the huge cave bear—whose Latin name I am not about to look up—was most likely a pure vegetarian judging by the cusps of its teeth.

Personally, I doubt this idea of fear of bears stemming from the very old days. It just seems common sense to me to be scared motherless of any animal that has the potential for carnage that any full-grown bear does. As far as I'm concerned, if you're not afraid of bears you're doing something wrong.

With the concept firmly in mind that we have to start somewhere, why not begin this horror-fest with the bear the psychoanalysts would have a ball with: *Ursus horribilis?* Really, now, what sort of behavior would you bloody well expect from an animal named, in formal Latin, no less, the "Horrible Bear?" Perhaps it's one of those chicken-or-the-egg things, but I suppose whether the grizzly is a most terrifying man-eater when so prompted or became such just to live up to his name isn't very important, particularly since he most likely cannot read and understands not a syllable of Latin. Sure, we're kidding around, whistling in the graveyard on our way home through the dark, moonless, bear-filled night. But there have been many times recently when there was nothing funny about spending the night in Montana's Glacier National Park. . . .

The August evening in 1967 was marvelous camping weather although the 60 glaciers and more than 200 lakes in the near 1,600 square miles of Montana wilderness probably had little to do with the coolness. A group of young people were bedded down in an area known as Granite Park, all asleep by midnight. Most probably they were having a nightmare when a bloodcurdling scream from 19-year-old Julie Helgerson raped the stillness, the teenager a little way off from the main body of sleepers. In the light of the anemic campfire, she was starkly outlined in the jaws of a trememdous grizzly bear. As she screamed and fought, it appeared for a moment that she might escape as the bear dropped her to severely bite a young man of the group in the legs and back. But, the bear seemed to prefer the more tender Julie and returned to bite her through the body and drag her several hundred feet where, for some unknown reason, he suddenly left the dying girl and ambled off into the night. By the time rangers from the park arrived, she was just a statistic. The young man who was also mauled survived after hospitalization.

It was a very bad night for 19-year-old girls in Glacier National Park. Just a few hours later, at four in the morning and 20 miles away, another group of campers were frightened awake by a grizzly who towered over them, growling like a thunderstorm. Like a flushing covey of quail, girls and boys scampered up trees and scattered into the blackness. All but one. Michele Koons of San Diego experienced the unspeakable terror of the zipper sticking on her sleeping bag! With the bear only feet away, she was bound and trussed by the unyielding nylon skin. The grizzly grabbed her. In numb panic, the rest of the party listened to her describe her own death. "He's tearing my arm off!" was one shriek all agreed upon. "Oh, my God, I'm dead!"

Right she was.

If either of the girls killed that night in 1967 was missing any flesh, the press and my sources did not mention it. I cite the night of terror as a precedent of attack rather than of man-eating, although one must wonder what the bears would have done had they come upon solitary campers and were not disturbed with their kills. The odds against something as rare as fatal grizzly attacks upon two girls the same age, in the same park, 20 miles apart on the same night actually happening have given me a rather eerie feeling when I read the monthly Solunar Tables created and copywrited by my old friend John Alden Knight, which still appear in *FIELD & STREAM*. These tables are purported to forecast periods of peak feeding activity for fish and game. That night they were pretty accurate.

The two grizzlies were shot and killed, proof of their perfidy having been confirmed by blood samples found on claws and muzzles.

We've discussed in the first chapter the problem of the new class of man-eater, the "park killer" so that does not bear repetition here but to reiterate that the bear is the classic North American example of this syndrome. Thirty-six people were mauled in less than 20 years in parks by bears and one more killed in 1972 at Yellowstone by a grizzly. Considering that well over two hundred million people (most are repeats) enter national parks each year, this isn't much of a toll. But, don't forget, not many parks have grizzlies.

It was nine years later that the first substantiated case of man-eating, or, if we carefully note the preferences of the Glacier Park bears, woman-eating occurred. It was another college girl, Mary Pat Mahoney of Highwood, Illinois, a student at the University of Montana. Mary Pat was 22 years old. She would get no older.

Camped with four friends, all female, Mary Pat Mahoney's tent was torn open shortly after dawn, and the girl, still in her sleeping bag, dragged away under the ripping, yellow fangs. Her friends, awakened by Mary Pat's screams of terror and agony, attracted the attention of another camper, who ran to get ranger Fred Reese. Reese arrived a few minutes later where he was joined by another ranger, a Californian on a "busman's holiday" named Stuart Macy. Just outside the shredded tent lay the gore-smeared sleeping bag and nearby, a bloody tee shirt. A clear spoor of blood and drag marks led off into a thicket, the partially-eaten body of what used to be Mary Pat Mahoney found about 300 yards from the site of her probable death.

Fred Reese, half-gagging at the sight, gave his .357 Magnum revolver to Stuart Macy who agreed to stand guard over the remains in case of the bear's return. Reese went for help. No sooner was he out of sight than a grizzly lumbered up and informed Macy that his presence was not appreciated. To top things off, the .357 was either defective or broken, which might be just as well as I, for one, have no interest whatever in putting any close range handgun bullets into any man-eating grizzlies while standing over their kills. Unless he'd gotten lucky with a brain or spine shot, Macy might well have found himself joining Mary Pat. As it was, the bear was sufficiently nasty and Macy had to climb for his life. His shouts and yells brought armed ranger help, two men with 12 gauge shotguns stuffed with rifled slugs. The first shot floored the bear, but, true to grizzly tradition, it got up and took off. Shortly thereafter, one of the rangers was able to pick out the form of a grizzly's head and blew a big hole in it. As it turned out, there were *two* bears, probably siblings, and by the human blood identified on both, they undoubtedly shared breakfast with the body of Mary Pat Mahoney.

A Board of Inquiry was already established after the 1967 debacle, but it could not determine that the girls had done anything to provoke the attack. They had even made a point of not bringing any meat on the trip to avoid bears! They wore no perfume and were in no known way provocative. I do not wish to be in any way indelicate, but I wonder, since so many victims have been female, whether the key factor could be menstruation and the detection of such by a bear?

That the problem is not improving, in fact is eroding into rank man-eating by grizzlies, was proved three times just in 1980 in good old reliable Glacier National Park. On the night of July 24th (and the age factor is starting to get spooky) a grizzly tore into a tent occupied by a young man and a young woman, both 19, (who may have attracted it by the scent and sound of doing what came naturally, although I do not know their relationship). Employees of McDonald Lake Lodge, the two were killed by the bear and the young lady largely eaten.

In October, it happened again. On the 3rd, the mutilated and partly eaten body of a Texas backpacker was also discovered near his camp at Elizabeth Lake (Glacier Park) close on the Canadian border. I wonder what the next few years will bring. I've got a pretty good hunch.

The grizzly is such an impressive carnivore that I am tempted to extoll his qualifications in a literary context where this is perhaps not warranted. He may weigh as much as 1,000 pounds, is in big trouble as the world crushes in around him and is the central figure of some of the greatest legends of the American West. He shares a well-earned reputation with the Cape buffalo and sand dunes in general for talent in the field of lead absorption, a characteristic learned by all on the Lewis and Clark Expedition. When he turns rogue, he's a national migraine. One, sporting the title

Old Mose, was at last killed in 1904 after having eaten more than 800 head of cattle and killed five men! That must have been a lot of bear!

Another terror, after years of raiding the sheep in the Wasatch Mountains of northern Utah in the 1920's (near the same place where in the winter of 1846–47 the Donner party had been trapped and resorted to cannibalism) was finally killed in 1923 by a sheep rancher and hunter named Frank Clark, who out-foxed the phantom bear with a second or "sucker" trap which sank its teeth into the bear's forearm. After dodging each other through a very nasty night, the bear, christened by the local Mormons as "Old Ephraim," charged Clark, who was armed with only a .25-35 caliber Winchester lever action rifle. (For you gun buffs, this was wrongly called a ".25-.35" in F.M. Young's account of Old Ephraim. It was a "nothing" cartridge in modern terms, firing a light 117-grain factory load bullet which has been called by the expert ballistician Frank C. Barnes, ". . . just about the minimum that should ever be used on deer, and in fact it won't qualify for this purpose in many states.") With the help of his little dog as a distraction, though, Frank Clark, after having twice shot the bear *through the heart,* stopped Old Ephraim with his last and seventh bullet at six feet, a brain shot in the ear. He was an inch short of ten feet long and held in such respect by the ranchers that he was actually granted a grave and buried, a suitable marker erected on the site:

"HERE LIES OLD EPHRAIM.
HE GAVE FRANK CLARK A GOOD SCARE."

You can count on that!

For no particular reason, let's proceed at this point to a perusal of the great bear of the north, the animal sometimes considered the only predator on the North American continent that hunts, kills and eats man as part of his normal prey. A pal of mine who has shot several, over a few beers the other day, put it quite well when he observed: "Anything on that ice is food for the polar bear. He's at the top of the stack except maybe for the killer whale, and looks at any other animal he can catch as dinner." I suspect he's correct, at least as far as a consideration of the polar bear in an unaltered habitat, shared only with the primitively armed Eskimo is concerned. But, then, with bears, you never know.

My fellow Editor at *OUTDOOR LIFE,* Ben East, makes a very good point in his excellent book, *BEARS* (Outdoor Life, Crown Publishers, New York, 1977) when he reminds his readers that Fred Bear, president of the archery tackle firm of that name and one of the greatest bowhunters of dangerous game in history, killed both grizzly and Kodiak bears without so much as a severe threat of a charge, but the first two polar bears he punctured with his razorhead trademark arrowheads charged Fred and his guide without a second thought. The men were lucky that both bears were able to be stopped at such close range with the guide's rifle, but neither counted as a bowkill because of the firearm interference. Fred Bear did eventually kill a polar bear according to the rules, but he still is convinced that that chap up north with all the yellowish-white hair is unquestionably the most dangerous. That is not an amateur opinion, either.

So far as I can determine, being a tropical bird by persuasion, and claiming no familiarity with the beastie, the polar bear is the one member of the species that is strictly carnivorous, or at least this is the case for the great majority of the time. Whether or not he eats french-fried tundra at certain times of the year doesn't really interest me, but I think it safe to say he'd probably die of starvation chained to a salad bar.

There is no question whatever that both polar bears and Eskimos spend huge parts of their spare time hunting and eating each other, which seems to be a nice, clear-cut relationship in an otherwise muddy world. As the object of some pretty shabby "sportsmanship" I know several people—I shall not dignify them as either gentlemen or sportsmen-hunters—who obtained their polar bears from the gun rest of an ice-cutter's gunwale off the northern Norwegian Islands. My good friend, David Putnam, son of the well-known publisher and stepson of Amelia Earhart, was part of a polar expedition that roped a female and two cubs under identical circumstances, so shooting a white bear in the water from a ship would be sheer murder. These are the same people who have wolf skins obtained on a trapping license and shot from the air with buckshot. Please, do *not* be tempted to confuse them with legitimate hunters any more than you would lump a jewel thief with a diamond cutter on the basis that they are involved with the same commodity. The use of aircraft in hunting polar bear, although once completely legal, seems outwardly a rather obvious means to an end, yet may not be so. I have not done it and believe it would be unfair to draw any conclusions without tasting the wine. I know too many absolutely ethical people who have hunted in this generally misunderstood mode to believe that it was unsporting. Apparently, the public misconception is that the aircraft is used as a part of the actual hunt itself. From my understanding, it is simply a vehicle to make access to the ice floes possible. A point well considered is the unbesmirched rule of the Boone and Crockett Club, which rules on and records North American game trophies, that the direct use of an aircraft to locate an animal is unethical and any trophy thus obtained will be disqualified as not consistent with the rules of "fair chase."

Well, if you're going to digress, do it properly. . . .

Thalarctos maritimus, whose name I know you were pining to know, is the second biggest of the bears after the Kodiak, which we'll get into in a moment. This position of supreme predator of his world of the Arctic, matched with his physical characteristics as well as the very low population density of people in his range tally up to a most uninhibited and effective general predator. By "general" I mean that man is fair game.

There is, considering how many more seals there are than Eskimos, a pretty reasonable school of thought that bears stalk and eat people because they think they are just another kind of seal. In afterthought, it's a fair notion, in my opinion, because the Arctic aborigines largely dress in sealskin, smear themselves with one form or another of blubber and are about the same size as some of the most populous species of seal. Eskimo hunters commonly, after spotting a polar bear, will lie down in a position a sleeping seal would take to draw the bear into range.

The Barents' Expedition in the 16th Century recorded many attacks by polar bears and at least one witnessed case of a member of the company being eaten.

It has been pointed out that perhaps polar bears are, as a group, rather baffled about just what the hell they have ahold of when they catch a man. Some authorities say that a bear normally eats only the blubber (fat) of a seal, discarding the meat, and upon finding none on a man, show clear confusion. Surely, polar bears in the wild run across precious few humans and fewer live to pass the information along. Plain curiosity may be the reason for such persistent digging-up of Eskimo graves, the bodies exhumed, not eaten. So, precisely why the white bear is a confirmed man-eater may not be so important as just knowing that he is. Personally, I'm not awfully worried about the entire prospect.

Ah, the Alaskan brown. The Kodiak. Now, that's a bear's bear! For quite a few years there has been the same sort of controversy as used to exist around the leopard/panther and who was who and what was what. Is the Alaskan brownie just a super-grizzly or the griz a runt brownie? Does it really matter from a practical standpoint what the relationship is? Only over a couple of sundowners. The Kodiak Island bear is sufficiently bigger in the adult phase to be distinguishable from the grizzly or the Eurasian brown bear that I don't especially care if these are races of the same animal. They all bite.

Beyond this observation, I can find no *specific* instances of man-eating in the Alaska brown bear. For one thing, his range is most likely far too limited to give him much of a chance as well as being greatly underpopulated by man. Of course, many hunters and other outdoorsmen have been killed and injured by the Kodiak, but that's a different story, indicating clear provocation. So much for potentially the most logical candidate for a "disaster" novel. Excuse me, I have been informed that there has already been one, centered around a genetically thrown-back cave bear.

The last American bear that is clearly guilty of chowing down on man is one that most even knowledgeable people would not suspect: the black bear.

Now, sir, it just happens that I have some experience of *Euractos americanus*, acquired about 20 years back, and which to this day still influences my general jaundiced outlook concerning any close association with bears, generally or specifically. I was, in fact, the witness to a savage mauling which, although provoked, gave me a pretty good idea, by interpolation, what it would be like to incur the displeasure of a grizzly, brown or polar bear, let alone a big black.

I was grouse and woodcock hunting with my brother Tom at Loon Bay Lodge in New Brunswick, eastern Canada, probably about 1962. My guide and dog-handler was a young fellow named Sheldon something, with whom I duly wandered off one morning intent upon unspeakable atrocities to the transient flight of woodcock we all hoped had dropped in on the good moon the night before. Before we were even off the lodge's immediate property, I happened to look into a nearby evergreen and spotted a black blob that could be nothing but a bear, and not much of a bear at that.

Well, Sheldon reckoned that was just dandy, as he'd been looking for a cub to raise and tame for, lo, these many years and this was clearly his chance. Handing me a couple of buckshot shells to cover him in case mama, who had apparently gotten a better offer elsewhere, showed up, he started climbing the tree with the serious intention of catching the cub, which couldn't have weighed more than about 40 pounds with a full stomach.

I can capsulize this whole adventure by saying that Sheldon is one lucky guide that he isn't chained up outside some hollow tree to this day! Man! What that little dear didn't do to him you wouldn't find in a "How-To" karate manual. By the time the bear ran off and I regained my strength lost through laughter and was able to stand again, Sheldon looked like he'd been covering the south wall of the Alamo all by himself. He'd been bitten three times, scratched like he'd been hand-sorting wildcats and practically broken his neck (and mine) when the bear dropped smack on his head and knocked him 20 feet out of the tree, bouncing off branches the whole way. I think Sheldon took up the breeding of tropical fish after that morning. Guppies, if I'm not mistaken. . . .

In proper, tradition-steeped bear hunter's terms, there is one whole passel of black bear in these here Yewnited States. Of course, nothing like there used to be, when Seton recorded presumably reliable sources as having killed 11 in one day and as many as 18,000 black bear skins sold by a fur company in a single year. Of the 50 States, Ben East says only a dozen do not have blackies. Alaska, like most everything else, has the most at about 50,000 and there are possibly as many as a quarter million or more conservatively in North America. I can assure you that we have no shortage here in Florida. The guide I used to fish with in the Ten Thousand Islands one morning showed me why we couldn't go out on our usual day. A huge bear had torn his boat apart trying to get the bait out of the livewells just east of Remuda

Ranch on U.S. Route 41, the Tamiami Trail, about 25 miles from Naples, Florida.

The few local Seminole Indians with whom I have signed a peace treaty—the United States Government never did—do not opine to be over-fond of the species, either. Actually, the Everglades bears are among the most respected in size, some openly represented as better than 600 pounds, which, when interpolated from the Florida pound to the American pound, would mean something over 350. Don't forget, we're the boys who brought you Ol' Slewfoot!

There are at least seven people recorded who find none of these goings-on in the least humorous, if, indeed, you do yourself. They were all killed and eaten by American black bears. But, before I specify these cases, perhaps a short heart-to-heart would be in order as my approach to the literary side of death has been referred to by some as less than, well, reverent.

I have seen quite a fine assortment of death, some of which I have been responsible for personally. It is my belief that those instances were morally justified either through self-defense, natural adaptation as a qualified predator or indirectly as a taxpayer. My philosophy is identical to that of Woody Allen who once wrote a line that well illustrates my view of life: "I don't want to achieve immortality through my work. I want to achieve immortality through not dying."

Not having any wish to type an inconsistent tumor into a chapter about bears, wolves and, let me see . . . hyenas, I shall state my case without benefit of soliloquy by observing that death is the only thing in the world you can absolutely count on. Not even taxes are so reliable, as we see shaping up under the new Reagan administration as I write this. I resent any formalcy, be it a religion or a sanitation law or the perfidy of the rare advantage-seeking funeral director, who mystically, sanitarily or economically places legal or moral brackets around my final physical rite of passage in which we must all return our salts to the earth. It is my personal view that death deserves no more respect as a human activity than any other; which means, not much. Besides, nobody ever gets very good at it. There's nothing sacred about dying; mystical, maybe. But that's only because we don't understand some of the more involved implications. My personal view is that we *do* have it rather figured out but don't want to face facts. We have an odd tendency to somewhat overestimate ourselves as a species, which insinuates colossal nerve from our kind who have been here any eye-blink compared to the dinosaurs' tenure. I am writing this book, for reasons beside the money, because I find the whole idea of animals that eat people interesting. Not holy, not bizarre, not even horrid (unless I might be personally involved). Yes, I feel sorrow for the youngster maimed or killed by an animal. Yes, I wish there was no cancer. No, I did not send in my Publisher's Clearing House Sweepstakes Entry Form. Not even the *READER'S DIGEST* sweepstakes form. You see, I'm a pessimist: something will have eaten me before I could possibly have

(105)

a chance to win. Somehow I just knew you'd want to know my philosophy. Back to bears. And death.

It was May of 1906, not an especially auspicious date for the first recorded case of man-eating by a black bear in North America, and most likely was anything but the original instance. It's just that folks, including Indians, probably had better things to do than write up such matters until this point. Three men were at a lumber camp on the Red Deer River of Alberta in May; two workers by the names of McIntosh and Heffern as well as the cook, a man named Wilson. The woodcutters saw a bear wander out of the woods across the river and shouted for Wilson to come out and have a look.

That bear must have either been starving or gifted with tremendous powers of resolution because, without a hitch, he walked through the river, stopped to shake himself off and without the slightest hesitation charged straight at the astonished group of three men.

The axemen made it safely to the cookhouse about 10 yards away, but the cook, Wilson, was leading a losing race with the black bear. The incident was probably funny to everybody but poor Wilson, who, in his enthusiasm to remain uneaten, couldn't slow up sufficiently to get into the cookhouse door. Around and around the shanty went the man and the bear, the black gaining at every yard. With a probably merciful swat, he reared up and broke Wilson's neck with one blow. The man was dead before he hit the ground, at least we hope so.

You'd reckon we were talking about a hardened man-eating lion rather than a lousy black bear, but this animal wasn't about to be driven off his kill. Both McIntosh and Heffern were close enough to hit the animal with, among other items, a square bullseye with a can of lard and the man-eater even ignored a perfect hit with a nasty canthook. He picked up the hopefully dead and certainly unconscious Wilson's body and dragged it a few yards away, giving the other two men in the shanty a chance to run like hell for the bunkhouse where one of them had a revolver. Perhaps they should have thrown it at the bear, based upon their accuracy with canthooks and lard cans, because despite repeated shots, no bear hair flew. Probably tired of the harassment, the man-eater carried Wilson about a hundred yards into the bush, out of range. It is insinuated that he had the chance to eat at least part of Wilson before realizing that discretion was the better part of man-eating when a mediocre rifleman arrived on the scene and demonstrated—unsuccessfully—the error of the bear's ways. Ah, well, style isn't everything: it's results that count.

I wish, at least from my side of the typewriter, that I could give you some good, hairy tales about man-eating black bears that didn't sound like overwritten "What I Did On My Summer Vacation" reports, but the black bear just doesn't have a great deal of thespian presence. In 1961, he ate an Ohioan (which, judging by some of my best friends can be hard to swallow at the best of times) in Ontario—which must violate some sort of international law—and earlier, in 1924, had solved all the problems of a trapper by the name of Waino, also in Ontario. The Waino case, however, reads more like Jim Corbett wrote it as the bear had had a losing brush with a porcupine that had given him both barrels in the face and neck. Although genuine man-eating, the Waino case would have to be considered in the light of a starving, injured bear. I'm sure Mr. Waino would feel much better about that.

A distinguishingly nasty incident that cost the life and much of the body of a tot, three-year old Carol Ann Pomerankey, happened in America's Upper Peninsula of Michigan in 1948 when the little girl, daughter of a forest ranger, was fatally bitten in the neck by a black bear. The bear had come out of the woods where the family was living in the Marquette National Forest, while Carol Ann's dad was at work and her mother involved in the kitchen. The little girl had actually reached the screen door of the cabin under the full view of the mother when the bear executed her. Despite the valiant efforts of the distraught mother to beat the animal off with a broom, it stuck with the body and carried it off.

A posse was formed within a short time, complete with dogs, and the bear was driven away from the body where it had stopped to eat part of the little girl. One of the hunters, a professional fisherman by the name of Weston, volunteered to stay with the remains while the rest of the men went on. Five minutes after the main group had left, Weston practically filled his pants when he turned around and saw the man-eater, standing on its hind legs, not more than 20 feet away. Terrified to chance firing and more terrified not to, he shot the bear square in the chops and then finished it with four more bullets in the body. It wasn't a very big bear, sort of a half-pint, and official examination of the animal's body showed no possible reason for the attack. It might be noted, though, that it was a lousy year for blueberries, a common denominator when black bears take to eating people.

Here's to good blueberry harvests. . . .

Okay, we've clearly established that American black bear is a people-eater and, when determined, pretty good at it. There has, though, been no case of a bear turning genuine multiple man-eater as the big cats do, so, beyond making the irrefutable point that the black bear does kill and eat man, I think we ought to let the poor chap alone.

There's a weird-looking denizen of India and other less pronounceable places called the sloth bear (*Melursus ursinus*) that everybody who writes about bears seems suspiciously eager to grant the crown for Man-eater of the Year. Were it not for the fact that outdoor writers in general and those specifically who scribble about man-eating animals are conspicuous by their obvious veracity, clear eyes (no smoking permitted in the library) and look of eagles, one might be remotely questioning

(107)

about the true ferocity of the *Melursus ursinus*. Ah, but having read this far I know you will agree.

Actually, no kidding, I once did an interview with a man who killed a man-eating Indian sloth bear. If you don't know the name Berry Boswell Brooks, you don't own a copy of Rowald Ward's *RECORDS OF BIG GAME*, in which he is prominently featured opposite many of the largest whatevers collected. I just may one day do a book about Berry, who was a character out of a mold that has been somehow misplaced, but since *bwana* Brooks was not himself a man-eater, I feel editorially bound to dispense with further dissertations on his character and hospitality. Berry got a better offer from above a few years back, but I still have the tapes of three days of interviewing him. By God, sir, he was the most willing interviewee I ever ran up against! As I recall, he even bought me some extra tape cassettes in case I ran low! . . .

Whatever the case, the sloth bear is the last one you're going to pry out of me. Rather like a two-headed toad, it's interesting in that it apparently, judging from its constant irritability, suffers from persistent, perhaps terminal constipation. The one Berry shot had killed an even dozen women, the remains of one of which was the bait he sat up over to place a .458 caliber hole through the killer from a .460 Weatherby Magnum. (Don't let Roy Weatherby, ultra-velocity gunmaker, confuse you; he just adds the sales tax to the caliber.)

One of the more interesting aspects of the sloth bear, who is no larger than an average, underprivileged American black bear, is that he tends to kill with his three-inch claws, biting only as a secondary method of attack. This brings to mind a parallel of another very badly confused Asian animal, the Indian rhinoceros. This poor chap generally bites anybody he can catch up with rather than goring as is more customary among his clan. Well, not to worry. For sure, upon reading these words, one or another of the "preservationist" groups, will get up a massive program to teach Indian sloth bears how to bite and Indian rhinos how to gore. If you're interested and they can get the government equally so, you will be proud to know that they may do it with *your* tax dollars, too.

*   *   *

If you have not already suspected such to be the case, there is no animal's supposed man-eating habits which encompass such a great degree of controversy and hard feelings between people of different emotional persuasion as those of the wolf, *Canis lupus*. The only indisputably true thing that can be said is that the wolf, at this moment, is at the very edge of a grinding, cutting, violent conservational tectonic plate, which increasingly shakes even the most emotional aspects of the study of animal behaviorism.

Perhaps you've noticed how many more books there have been published recently on wolves, particularly the rearing and keeping of them in captivity or semiconfinement, sometimes in a *BORN FREE* motif of returning the animal to the wild. In fact, although I draw no special conclusion because of its insertion beyond subliminal, one of the pictures in Caras's chapter on wolves is captioned: "The submissive posture of a female wolf in a highly socialized pack. The woman is Joy Adamson, author of *BORN FREE* founder of the "Elsa Wild Animal Appeal." This seems to imply that Ms. Adamson, who is petting or touching the wolf, approves of wolves, particularly because she is smiling broadly. I'm not sure what the message of the picture is supposed to be, but it would take a flamingly vivid imagination to believe that cringing and submissive lady wolf is any tougher than a ten-year-old mink coat.

I'll give you one piece of good advice: if you happen to take a stance other than that of the New Wolfmen, you might be well advised to stay off talk shows with them. They're not exactly the most logical of reasoners and can get very excited about what a bad guy man has been to the wolf over the years. Further, they would be mostly completely correct.

Since there is so much controversy around the "goodness" or "savagery" of the wolf, I can't see any approach to this subject of man-eating other than purely statistical and base any conclusions we do reach purely upon the most reliable records we can dig up. As in most matters controversial, my guess is that the truth will be found, cringing like a wounded rabbit, somewhere pretty close to dead center between views.

To give you a practical idea of what we're up against, the first thing that James Clarke tells us about wolves is that, although it's mighty strange, the wolf has never been known to eat a man in North America. Yet, John Pollard, who wrote *WOLVES AND WERE-WOLVES* (John Hale, London, 1966) a pretty definitive book about European wolves sees the beast as sufficiently "cruel" to justify the old and quaint custom of sewing up the lips of captured specimens and then having them skinned alive! Good Lord, but is *anybody* being remotely objective?

Although I promised to keep this subject as free of personal conjecture as possible, one facet of the human-wolf relationship is undeniable: it is ancient, going back as far as we do. We have always been impressed by the efficiency of the wolf as a predator, which is understandable as that's what we're in business for ourselves. Since the development of firearms and a much more active competition for prey animals, man has beaten up on the wolf undeniably, basing his relatively arbitrary policy of emnity upon a variety of both valid and invalid excuses. There's no black and white with wolves *or* people. Sure as shootin' wolves kill stock. One, the legendary Custer wolf, who was killed in 1920, had a bounty of $500 on his head and was, interestingly, accompanied by two coyotes during most of his ten years of devastation in South Dakota and Wyoming. Some estimates reckon he killed $25,000 worth of stock. Twenty-five grand bought a lot of lamb chops in 1920, too.

Probably because the wolf is good at what he does and kills as a group, man got an early impression that, if he was of such a mind, the wolf, because of his obvious intelligence, would not only be a potential problem but clearly represented the forces of evil. I'm sure howling at night while early man sat around a dying fire, trying to protect his newly domesticated stock didn't help the wolf's popularity either. Thus, just as the Africans have their traditional bad actors such as leopard-men and werehyenas and werelions, we have the werewolf, the transmutation of the most crafty and intelligent predator we recognized. The idea was so popular that it hangs on today as Little Red Riding Hood, the sexual "wolf," the "wolf whistle," to "wolf down" food, "the big bad wolf" etc. Wolves have a bad image among the average Joe, deserved or not.

In Medieval Europe, there may be pretty good documentation that the reputation of the wolf as a man-eater was justified, too. When Paris was still a walled city, in 1447, 45 years before Columbus got lucky, there was an infamous bobtailed wolf called Courtaud, who led a pack of about a dozen other wolves and terrorized the suburbs all summer and fall. That winter, Courtaud and his gang are reported, reliably or not, to have gotten through the wall and eaten 40-odd people before being baited to the square in front of Notre Dame Cathedral where they were speared and stoned to death. True? Probably at least to some degree.

There was sure no question about the "Beast of Gevaudan," which Pollard records as a single animal. Barry Holstun Lopez, quoting a C.H.D. Clarke, a Canadian naturalist (whose original material I cannot find in Lopez's bibliography of *OF WOLVES AND MEN*; Charles Scribner's Sons, New York, 1978) maintains that clearly there were two animals involved, ergo, Clarke is credited with bringing the "Beasts of Gevaudan" to an English speaking audience.

Fortunately, there's a good cross-reference to this version in *THE WORLD OF THE WOLF* by Russell J. Rutter and Douglas H. Pimlott (J.B. Lippincott Company, Philadelphia and New York, 1968) which reveals Clarke to be a Doctorate holder with the Ontario government. The manuscript Clarke wrote was never published and is undated. Because of physical details of both wolves as recorded by an original research source, Abbe Francois Fabre, in 1901, from old parish records and such I'll buy it that there were two wolves instead of Pollard's version of only one man-eater of Gevaudan.

The story begins in June of 1764 on the central plateau of France, which Pollard figures had more wolf casualties than any country in the world. A single wolf rushed a woman herding her cattle but she was uninjured as the animals drove the wolf away. But, then, on July 3rd, a teenage girl was dismembered and eaten, best guesses indicated by two wolves. By the time October rolled around, ten children were eaten and six more before New Year's day. The fact that three people were eaten over the holiday and three more on the 6th and 7th of January seems pretty conclusive that more than one wolf was doing the eating or he would have developed heart problems from being so overweight. By the end of 1765, despite tremendous hunting pressure and huge drives, the death toll had risen to 50 persons.

The single animal version of the Beast has him killed on June 19, 1766 by a 60-year-old farmer named Jean Chastel, one of more than 300 hunters in the La Tenaziere forest, the wolf having totalled 60 kills. In the two-wolf version, the first animal was killed in September of 1765 and the second in June of 1767. By any standard, they were very large, one recorded at 130 pounds and the other at 109. Because of a comment one writer makes, at least the skull of one of the Beasts must survive as it has been measured. Dr. Clarke is reported to have wondered in his "phanton" manuscript if, through description, they were not wolves at all but hybrids with dogs. Whatever they were, they killed at least 64 people and attacked about 100 overall.

The European and North American wolves are the same animal, which makes for an interesting question as to why the European variety seem to have such a record of human depredation whereas it is very difficult to pin down a genuine case of man-eating in North America, let alone a determined attack. Of course there are dozens of races of *Canis lupus,* but there are all variations on the same theme, with one exception, that of the Red wolf group, which are members of the *Canis* family but not of the *lupus* group, being of the *niger* classification. So far as we're concerned, this discussion only extends to *lupus,* no matter what the race.

Some views have been expressed that the European wolf has the history of attacking because of the much longer association with man through the relatively late discovery and colonization of America. This doesn't seem to hold water, under scrutiny. Nowhere can I find a strong tradition of man-eating by wolves among the American Indian, although the joint must have been crawling with wolves when the bison was in his heyday. There are plenty of reports of wolf attack by settlers, but these traditionally are fanciful and almost none will withstand investigation. Another to ponder is that European wolves, even in very recent times since World War II, and probably even today are still eating people in Europe and Asia!

World War II was manna from Heaven for the wolf population of Europe. All the hunters were off fighting each other and the glut of abandoned corpses in the wilder places (imagine how many bodies the combined Germans and Russians left unburied during the Russian Campaign?) insured that the weakest of cubs would survive. After the war, particularly in the north, where there was less habitat disruption, a lot of problems developed when the war ended and all that easy food disappeared. On the border between Poland and Rumania in 1946, a pack of more than 50 wolves ate two soldiers before they were chased off with automatic fire and grenades. Finland was having such a problem by 1949 that it had a whole campaign mounted against wolves which had been eating both stock and people. Despite the use of aircraft, machineguns and even mines, only a few wolves were killed, the rest coming back out of the deep forest as soon as the pressure was off. Eleven children were eaten by wolves in Portugal (Portugal!) in 1945. In ten years, reported only up to 1955, there were just short of 7,000 wolves killed in southern Yugoslavia. Although the wolf is probably no longer found in its old stronghold of France, a schoolteacher was eaten in her own classroom in 1900 and the last reported death in that country was in Dordogne, in 1914, where so long ago poor Neanderthal probably was cooked and eaten by Cro-Magnon. She was an eight-year-old girl eaten near her house.

Spain has always been rife with wolves and still is in the wilder areas. Italy also has very viable wolf packs, including recent man-killers. A postman was killed and eaten about two hours drive from Rome in 1956; you read right, not 1856! Six years earlier a soldier was torn apart after defending himself with his bayonet, killing one wolf before the rest got him.

The vastness of the Russian steppes and taiga have always been ideal wolf habitat, and man-eating has never been rare. The village of Pilovo, according to James Clarke, was damned near wiped out during a food shortage by a concerted wolf attack in 1927, the survivors under siege conditions until rescued by the army. The cheery Pollard tells the apparently true story of a caravan being completely wiped out by man-eating wolves in the Ural Mountains about 1914. The Russians reported 30,000 wolves shot in a single year in the 1960's and more than 70,000 assorted barnyard animals killed by them, plus 11 people eaten out of 168 reported attacks on man.

One of the most bloodcurdling aspects of wolf attack (not man-eating) and certainly one of the most common causes, is that of rabies. In fact, it is almost unquestionably the single greatest factor. Wolves being social animals, rabies can spread quickly through a pack. An entire section would be devoted to this aspect of sheerly horrible behavior, but we had better stick on the tracks which lead from man-eating to man-eating, at least for the sake of space.

Having clearly determined that the European wolf has a definite, certified history of man-eating, it would seem time to take another look west and see what the records show for North America. Was James Clarke right that there has never been a confirmed case in North America? Of actual man-eating—even if some attacks might have logically resulted in the losing man being consumed—there exists no evidence I can consider "hard" and I've turned over one hell of a lot of reference rocks. Of course, some of this becomes a matter of judgment. Elmer Keith, that great *Maestro* of the six-gun and other related matters, wrote to Roger Caras of knowledge of a man's skeleton being found on a island off southeast Alaska along with a weathered Smith and Wesson .357 Magnum revolver (which dates the incident to no earlier than 1935, when that caliber was developed) and the associated skeletons of three wolves can only be considered circumstantial. There is no proof that the man didn't attack the wolves, have a heart attack himself and get eaten after death by the crows. Perhaps he was simply killed by a wounded wolf or wolves. *Maybe* it was a genuine case of unprovoked man-eating, the trouble is that there are too many other possibilities, no matter what one prefers to read into the imagined "probabilities" of the evidence, it is just not conclusive.

Certainly, there have been reputable recorded instances of attack in Canada and the United States by wolves. A beauty was reported, complete with witnesses in a scientific journal of the Royal Ontario Museum, which includes the sworn testimony of one Mike Dusiak, the victim, the train crew that battled and finally killed the wolf and the wildlife official who investigated the incident. Although it reads classically like a report of rabid behavior, nowhere in Dr. Randolph L. Peterson's paper, *A RECORD OF A TIMBER WOLF ATTACKING A MAN* (Journal of Mammology, 1947) is there any mention of rabies. However, as Rutter and Pimlott are quick to point out in their reproduction of the paper, rabies tests were not normally carried out at that time and, through extreme good fortune, Dusiak was either not bitten or did not contract the disease, the only basis for serious consideration that the attack was preliminary to possible consumption. Here's what happened, paraphrased from Peterson's rendition of Dusiak's testimony:

Mike Dusiak, a railway section foreman, was driving his "speeder" (I presume a slang term for a hand car or similar vehicle) slowly west of Chapleau, Ontario, expecting to meet a scheduled train. Suddenly "something" smashed into him and grabbed him by the left arm, hitting him hard enough to knock both him and the speeder off the tracks. (If it grabbed his arm that hard wouldn't the skin have been broken?) It was such a powerful blow he thought he'd been hit by a locomotive. Getting to his feet, he saw a wolf about 50 feet away, which immediately charged him. He grabbed a pair of axes, one in each hand, and laid open the wolf's belly with a stroke and lost one axe. The wolf then started a circling attack, so close that Dusiak was able to smack him in the skull with the second axe a couple of times, but seemingly with no effect. All the time, he was growling and gnashing his teeth and after what Dusiak thought was 15 minutes, he had been able to brain the wolf five times, each time he broke the circle and jumped at the man. Another ten minutes went by as Mike was able to keep the animal off, crossing the tracks and fighting until the train came along, stopped and the foreman got help from the engineer, fireman and brakeman, who together killed the persistent wolf with picks and shovels! The wolf was thought to be in excellent health, but was not tested.

Earlier, in 1926, there was a huge flap when what most authorities reckon was a rabid wolf walked into a small town of Churchill, Manitoba. As James Clarke reports, the army was called in to get the wolf, which was now missing, and within a short time most of the town's huskies and at least one local Indian had been shot. A tourist later ran over the wolf with his car, presumably an accident.

Those are the hard and rather remote facts about the wolf as a man-eater. In Europe, guilty as charged; elsewhere, no. I have looked long and hard for any other conclusion, but it doesn't seem warranted.

Why? Nobody knows, let alone agrees, but I have at least one idea. There's an interesting parallel between the wolf and the African hunting dog, *Lycaon pictus,* the incredibly efficient, highly socially-organized predator of the plains and savannahs of Africa. As rich in what would only be considered elaborate abstract ceremony and behavioral ritual, exactly as is the wolf, this truly fearsome killer, from which all game less than the elephant flee, is also one of the only predators of his continent capable of eating man who has never been reliably reported as doing so.

Now, that's more than just curious.

In my opinion, the reason the wolf is reticent as a man-eater when healthy and well-fed, even in Europe, is that he is a methodical hunter. He's not like a consistent man-eater such as the more solitary cats, who make quick decisions and attack targets of opportunity. The wolf hunt, from first howl to last bloody scrap is a masterpiece of organization. As such, I believe that the wolf recognizes us not only as a danger, but as another predator, not a prey animal. If you've ever seen the efficiency with which wolves kill, you'll join me in hoping that it stays that way!

*   *   *

If there is an animal as universally loathed, ridiculed and secretly feared through all the lands of those who know him, it is the hyena. As what may possibly be the most accomplished processor of human meat of all, both as a scavenger and newly recognized super-predator, old *Fisi* as most of the old East African types call him, has a very annoying habit that tends to rankle humans of every color: he always has the last laugh.

Hyenas are fascinating bloody animals. Despite the presumption that they are some sort of dog designed by a committee with their grinning, quite expressive faces and tremendous jaws, seemingly crippled, slanting downward hindquarters and one-man-band repertoire of the goddamndest collection of noises one ever did hear, they give a most erroneous impression of being some sort of animal fugitive from all that good fun on *THE GONG SHOW.* Spend a little time in the bush with *Fisi,* pal, and it won't be long, if you pay attention, that you'll respect him more than the lion.

It's interesting that, considering all the big, fat grants that have been financing African animal research for the past 60 years, that it's only quite recently that we have started to get away from the traditional concept of the spotted hyena, (*Crocuta crocuta*), as some olfactory offense to mankind, some "fringe" species of little value to anybody except for the processing of animal garbage and the entertainment of insomniac tourists. That, as an individual and small hunting band, the hyena may be the most successful predator of them all, in some places turning the tables so severely that the lions follow the hyenas and eat *their* scraps, as we've discovered is quite common, was not well known until just a few years ago, after the work of men like Dr. Hans Kruuk, and Hugo and Jane van Lawick-Goodall (before their current state of marital disenchantment).

But, let us trip lightly back onto the path of man-eating and the hyena. For purposes of space, let's consider just the most populous of the group, the spotted, and leave the fringe of the family out of it although the striped hyena does eat people in Asia.

The biggest problem with the hyena is the fact that he has a rude tendency to eat and run. If you spend some time in Africa, you will see some people around with faces that would cause you to gag up last Easter's jelly beans. As the late Robert Ruark, who I suspect strongly of having had a fondness like mine for hyenas, despite the way he reviled them in print, learned, there is quite often a relationship between those people with profiles that end abruptly below the eyeballs because of hyena bite and a fondness for the American equivalent of Old Stumpblower. Africans who drink themselves into the death-like stupor produced by some of this homemade liquid form of Roto-Rooter, sooner or later may be noticed as they are passed out near a dying fire. *Fisi,* albeit a bit prematurely, decides to sample what he's going to get anyway a few years down the road and SNAP! Look, Ma, no face! Depending upon individual circumstances, it may be no arm, foot, fingers or—let your imagination roam. Oh, yes. That happens, too.

Hyenas in general are brighter than some people I have worked for. At better than 150 pounds for a biggie, living around a clan is sort of like sharing an apartment with "The Munsters." I am convinced that, except for the primates, the hyena is the only animal I have seen who seems to have a sense of humor. It takes a while, but when you've spent years in Africa as I have, without TV and even sufficient light to read by on many nights—as if you would have had the strength—the hyena can become a frequent source of entertainment. I have recorded hyenas imitating lions in most convincing fashion, unless one happens to know the voices of all the resident lions well enough to differentiate. Really, I am positive the hyena will imitate lion roars, and I can't think of any reason other than for the hell of it.

I wish I could use the space to tell again my experiences in Botswana with my

(115)

hyena "retrievers" when sand grouse shooting, but you can find it in *DEATH IN THE LONG GRASS* (St. Martin's Press, New York, 1977). It'll point up the fact that free enterprise is not fading among the *crocuta!*

Unfortunately, my misplaced affection for the species is somewhat watered down upon realizing that, as a man-eater, he's probably a "natural" such as the crocodile. African custom has dictated abandonment of dead and dying people, leaving them for the hyenas so that the eating of people, dead or alive, has become ingrained.

A white by the name of Balestra killed two big man-eating hyenas who had ingested the village idiot of the Mlanje District of Malawi back in 1955. The killer hyena ate everything but the man's clothes, which I'm surprised were left behind as hyenas commonly eat even the bloodstained handles off knives, as they did to my faithful Randall skinner. A little after this, an old woman was taken bodily—and I do mean bodily—from her hut, but although a tribesman drove off the hyenas, her arm had been completely bitten off and carried away by the attackers. She died the next day of shock and additional bites around her throat. The next victim of the Mlanje hyenas was a six-year-old child, who had been killed with the ever popular technique of biting the child's face off. The hyenas had been joined by others, now, and ate the entire body but for the skull, half of which was left.

Between 36 and 60 people were eaten by this hyena group until 1961, all taken during the warm months when they were sleeping outside. Not too bright, guys! At last things died down, no pun intended. Balestra was never sure if there was *mbojo* or lycanthropic witchcraft mixed up in the matter as George Rushby found the case in Tanganyika.

I see little point to a long, drawn-out history of man-eating by hyenas, because it's not very entertaining except in a few cases I have already written about elsewhere. Suffice it to say that the hyena is a casual killer who will take whatever flesh he can get whenever an opportunity presents itself. The problem is, that includes you and me; although I promise you, everything else equal, it's a lot more likely to be you than me. You see, I've been there, Charlie!

LEON PARSON

CHAPTER SIX

TIGERS

On the basis that the great cats are pretty unquestionably the leaders of those terrestrial mammal forms that tend to snack on *Homo sapiens,* it follows that there's a pretty good body of opinion that the largest of these cats is probably the most successful as a man-eater. As far as the records indicate, this is true; or, it's true if the records themselves are true. The individual cat with the highest number of recorded human kills is the Champawat tigress who, after being chased out of Nepal in the first few years of this century continued its career in the Himalayan foothills of India, finished with the round, simple, easy-to-remember official tally of 200 Nepalese souls. Arriving in the Naini Tal area of Jim Corbett, the Champawat tigress fell at last to the borrowed bullet of that great corrective dietician of man-eating leopards and tigers with the rather untidy and sloppy final bag number of 436 kills, which must have irked the hell out of the neater, officious types all over the *Raj.* Point is, we'll never know whether or not this tigress actually killed more people than her closest competition, the Panar leopard, with whom officialdom had its way, sentenced to moulder away in the history books with a nice, concise credit of precisely 400 Indian hill people. Whatever the logistics, a grand total of 836 people shared between just two wild animals is still one hell of a lot of death!

Conjecture momentarily aside, there's no argument that a tiger turned to people-eating is one major flash of bad news. Personally, if forced to make a choice of having to try to sort out and kill A) a man-eating tiger; B) a man-eating leopard; C) a man-eating lion; or D) none of the above, beyond the obvious choice of the last category, I would far rather tackle the tiger. There is, in glaring contrast to most of my decision-making processes, some very solid logic behind this: what I know first-hand about wild tigers from personal experience would fit with great gaps of left-over space on a note of apology from Ivan the Terrible. I have never seen a wild tiger and probably never will. On the other hand, I have practically had lunch with far too many man-eating lions and leopards and they continue to scare the Holy Deuteronomy out of me. I would, on that basis, opt for the devil I don't know.

Practically every researcher involved with tigers discovers that the primary source of information on the subject of man-eating behavior in this species is Colonel Edward James Corbett, who wrote well and with the advantage of experience and perspective. Many books on the subject bear his authorship, including the immortal *MAN-EATERS OF KUMAON*. Well, except for acknowledging Corbett as the man who killed the two greatest man-eaters of all time, you may now go and pour a fresh brandy, snug in the knowledge that you have probably heard the last of *sahib* Corbett, at least as far as this chapter is concerned.

The tiger, that striped, sleek, sinuous symbol of the quiet, jungled Asian places, is the largest, strongest and unfortunately the most ecologically vulnerable of the race he tops, *Felidae*. Even a hundred years ago, before the internal combustion engine, he was so numerous that some experts, in retrospect, wonder today if there wasn't a point at which it was questionable whether man or the tiger would ultimately be the master of that sprawling continent. One authority quasi-documents that there may have been between a minimum of 300 and a maximum of 800 man-killing tigers operating at the same time just in India alone in the 1800s. Yet, as has been the cry of the anti-hunter far more interested in the sentiment his position represents than the facts of the matter, the question of the tiger is at least closer to the truth than that of any other species beyond the American bison, who was, unlike the tiger, not slaughtered by sportsmen but by official interests to break the economy of the Indian tribes. The truth about the decline of the tiger lies in three areas, two of them common contributors to wildlife downfall. First, habitat and the destruction thereof. Today, to create the perfect tiger sanctuary, an area of about 3,000 square miles would be needed—appropriately stocked with wild food—to accommodate about 300 breeding tigers. This would be roughly three times the size of Haiti. Of course, no people could be permitted to disrupt the arrangement with human habitation and climatic conditions would have to be right, offering fertile soil to provide vegetable food for prey. So, unless you know of a real estate agent who can put you on to a place with these requirements at a reasonable price in a starving world, you will see the hopelessness of trying to save the tiger in his natural surroundings in direct competition with man and his population explosion.

The second factor, although not in my opinion nearly as important as loss of habitat which seems to be primary to every species, is that of poaching and market hunting. As we have successfully proved and have had corroborated by the International Union for the Conservation of Nature and Natural Resources, the leopard has easily evaded poaching pressure throughout its range and has demonstrated an ability to adapt to constantly changing conditions for the worse in its environment. But, perhaps leopards are smarter than tigers, or maybe just smaller and less obvious. Tigers can weigh better than a quarter-ton and are considerably more highly profiled on the list of predators than the secretive, nocturnal leopard. In any case, the tiger has suffered terribly from *illegal* commercial pressure and fur collection which has led to drastic local population reductions.

Third, it is true—and important that the real sportsman realize that there were times, even recently, when such a thing could happen—that the tiger was over-shot for sport, albeit under governmental and social prestige conditions unlikely to exist elsewhere.

India is a strange and fascinating land, to a stranger a rather maggot-ridden enigma contrasting the very wealthy with the very needy in a degree rarely seen. The hard economic facts of life point up only too clearly that some people are smarter, luckier, stronger, more fortunately born or generally more successful than others, and *not* necessarily because they worked harder; keeping one's nose to the grindstone may produce either a bloody grindstone or a calloused nose, it rarely produces a princehood and a palace in India along with one's own stable of hunting elephants. Whatever the case, there were enough Indian nobles controlling enough land, local people and tiger hunting equipment, as well as having the spare time in which to indulge in the sport of tiger hunting to equate into bad news for poor old *Felis tigris*. Additionally, there were British troops in the country for generations and, along with pig-sticking and other forms of *shikar* (the sport of hunting, roughly translated) the most popular way for an officer to take his leave was to go tiger shooting. Between the Brits and the Maharajas, there's pretty good reason to believe that at least 100,000 tigers got the deep six over the past hundred years. (Personally, since this is only 1,000 per year, I can't see the harm.) Some researchers will tell you that this number was killed since the beginning of the 20th Century alone, and, when we look at the number of survivors through the haze of now-settling dust of the great *Terai*, they might be right. Through the incredibly expensive machine that was the Indian tiger hunting establishment, official guess-timates (now a couple of years old) are that, in India, of the some 40,000 left about 1930, less than 2,000 may yet exist. Sumatra has less than 1,000 left, and Java, once a real stronghold, probably has less tigers than one of us has fingers and toes.

(121)

Some of these hunts, especially those organized by the Indians themselves for British or other foreign dignitaries, were for sure not the equivalent of a fast-food operation. Take, for example, the comments of one Captain Thomas Williamson in his *ORIENTAL FIELD SPORTS* (Orme, presumably London, possibly Bombay, 1807). After 20 years in Bengal, which would bring his service back to at least the 1780s, Williamson describes a tiger "beat" or "drive" as using as many of 3,000 trained elephants, 30,000 to 40,000 (!) horses and unnumbered human beaters. The Indian nobility, who were the primary patrons of the emerging double-barreled British rifle trade, those most expensive of all heavy express rifles, encrusted with gold and engraving, (sometimes as many as six perfectly matched rifles being made to order for a particular potentate) had not yet the advantages of modern firearms. Still, bags of 30 or more tigers were not particularly rare in a single hunt.

Actually, consulting various journals and antique books, it becomes clearer where many of those tigers went: the Maharaja of Surguja in the late 1960's had personally killed 1,150 tigers by that date. One mere British Army major, hunting in his "spare time" in the State of Rajasthan, shot and killed or wounded over 150 tigers! Even King George V of Britain in a single *shikar* in 1911, personally put blue-edged .470 caliber dots between the stripes of an incredible 39 tigers.

Perhaps the last really big hooplah involving a member of the British Royal Family "coming out" for a state visit/tiger shoot was that of Queen Elizabeth II in 1961, who was a guest of the Government of Nepal. As might be expected, the Nepalese really pulled the plug!

The campsite for the exercise was reported as two miles square from which all insects and snakes were removed by hand and exterminated. Electric generators and even sprinkling systems were installed and the Queen, in the shade of a royal purple umbrella held over her head, led the hunting procession while riding behind the mahout of the lead elephant. In her not inconsiderable wake followed another entourage of 18 elephants and an undisclosed number of cars full of VIPs. The Queen, alas, was not a knock 'em, sock 'em tiger shot, but after several tries did manage to kill a smallish female, which monumentally irritated quite a few of the loyal bird watchers back home. Over the years, there has been, at least in my observation, in Britain a growing, and not very pleasant, feeling of shooting being more than simply a presumably horrid blood sport but a rite of the privileged; an occupation of the rich which has somehow clung to their socialist culture rather like some leftover conservative pterodactyl roosting bat-like in the national hall closet.

Of course, the equivalent of a full-dress rehearsal of the Battle of Waterloo was hardly necessary for the killing of a tiger or two. Among ordinary officers, a *Shikari* or Indian professional hunter was often employed under whose directions live baits, usually bullocks or buffalo calves, were tied out within shooting range of a tree

(122)

platform called a *machan*. Theoretically, the tiger or tigress would collect an impressive series of perforations when it came to kill the bait or when it returned to feed on the carcass of the bait later.

In the context of *machan* hunting possibly appearing unsporting to the uninitiated tiger hunter, it should be pointed out that contrary to most beliefs, tigers can and do climb well, although not frequently. In fact, they've been reported as far as 60 feet above the ground in a tree. One man-eater, somehow captured, was released years ago into an enclosed arena by some Maharaja who apparently could think of nothing of greater interest that particular afternoon. The blood of a lot of tiger hunting onlookers must have been chilled about proper for serving champagne as the tiger scampered without the slightest hesitation 30 feet up a smooth limbless tree trunk.

Tigers are also pretty fair at climbing elephants to get at the hunter, and many's the chaperoo for whom the regimental band turned out next day to solemnize his funeral. This nasty habit of tigers charging the hunting elephants and reaching the shooting party atop, led in earlier days to the custom of carrying those peculiar pairs of heavy caliber handguns called Howdah pistols, so named for the *howdah* or platform fixed atop the elephant in which the hunters rode.

The third method of basic tiger hunting was to combine the first two, driving the tiger with beaters and, perhaps, elephants past an ambush point, commonly a *machan*. The main difference in this bit was that it always took place in daylight, whereas, waiting up over a bait, dead or alive, was normally a nocturnal affair. Concerning this last method and the potential hairiness thereof, I cannot resist recounting the incredible adventure during a beat shared by a Mr. and Mrs. E. A. Smythies, around Christmas, 1925, in India's Haldwani Division. It so nicely reflects the flavor of the time that I believe it best quoted directly from the Bombay Natural History Society's *PIONEER* of January 30, 1926.

> "We were staying for Christmas in a good shooting block, and one night we had a kill by a tiger in one of the best small beats in the area. So my wife and I went off to the beat, and I fixed up two machans, my own in front, and hers about 40 yards to the right and behind, thus avoiding the risk of ricochets. Her machan was in the first fork of a tall cylindrical tree, 14 feet from the ground, the tree being 4 or 5 feet in girth. Just in front of my machan was a patch of heavy *narkal* grass about 25 yards in diameter, and there was a good deal of grass and undergrowth all round. Soon after, the beat started, and I heard a 'stop' clapping, and the tiger roared twice. About three

minutes later, I heard it coming through the *narkal* grass, and presently it broke cover at a fast slouch. My weapon was a H.V. .404 Jeffrey magazine rifle, with which I have killed several tigers. I had 4 cartridges in the magazine and chamber and several more loose on the machan. As the tiger broke cover, I fired and missed, whereupon he rushed back into the *narkal*. Presently the beat came up to the *narkal,* and almost simultaneously the tiger again broke cover, this time at a full gallop with a terrific roar. I fired at it going away on my left and again missed. The beast went by my wife's machan at a gallop about 30 yards from her, and as soon as it had passed her, she fired and hit about 6 inches or so above the heart and just below the spine. This stopped it, and it rolled over roaring.

"Here the incredible part of the story begins, The tiger, mad with rage, turned round, saw her in the machan, and made for her, climbing the tree for all the world like a huge domestic cat, with its forearms almost encircling it. Up it went vertically under her machan, and as I turned round hurriedly, I knocked the loose cartridges out of my machan to the ground. As things were, I had no option but to take the risk of hitting my wife. I fired at the brute when it was half-way up the tree, but only grazed it. As I looked to work the bolt and reload, I realized I had only one cartridge left, and, looking up again, saw my wife standing up in the machan with the muzzle of her rifle in the tiger's mouth—his teeth marks are 8 inches up the barrel— and he was holding on to the edge of the machan with his forepaws and chin. In this position she pulled the trigger—and had a misfire! You must realize that at least two-thirds of the tiger's weight was now on the machan, for, except for his back claws, he was hanging out from the tree by the width of the machan, which was rocking violently from his efforts to get on to it. The next thing I saw was my wife lose her balance and topple over backwards, on the side away from the tiger.

"The beast did not seem to notice her disappearance, and, as I again aimed at him, I saw him still clawing and biting the machan—the timber was almost bitten through, and the strings torn to shreds. I fired my last available cartridge, and, by the mercy of Heaven, the bullet went true. It took the tiger in the heart and he crashed over backwards on to the ground immediately below the machan, where he lay hidden from view in the grass. I did not know at the time that he was dead; nor of course did my wife. All I knew was that my wife had disappeared from the machan on one side of the tree and the tiger on the other, and that I had no cartridges left; and that I was helpless for the moment to give any further assistance.

"Whether my predicament was as bad as my wife's can be judged from her view of the incident. I quote her words:— 'When I fired again, he turned round and saw me, and immediately dashed, roaring, towards my tree. I thought he was galloping past, but suddenly realized that he was climbing up, and only just had time to stand up in the machan before his great striped face and paws appeared over the edge, and his blood and hot breath came up to me with his roaring. I pushed the barrel of my rifle into his mouth and pulled the trigger, but the rifle would not go off. Then I really did feel helpless and did not know what to do. We had a regular tussle with the rifle and then I saw his paw come up through the bottom of the machan, and, stepping back to avoid it, I must have stepped over the edge of the machan, for I felt myself falling. I thought I was falling into the jaws of the tiger and it flashed through my mind "Surely I am not going to be killed like this." I never felt hitting the ground at all and the next thing I knew was that I was running through grass and over fallen trees, wondering when the tiger would jump on me.

"She arrived at my tree almost simultaneously with the mahawat, Bisharat Ali, who had rushed up his elephant, regardless of wounded tigers or anything else, and she hastily mounted and cleared off into safety, unhurt except for a sprained wrist and various scratches and bruises from the fall. One of the 'stops' was calling that he could see the tiger and that it was lying dead under the machan. So, when a supply of cartridges arrived, I went up cautiously and verified his statement, recovered my wife's hat and rifle, and went off with her to the forest bungalow, leaving the 'stops' to bring in the tiger.

"It was a nice male 9 ft. 3 ins. in length with three bullets in it, one between the heart and spine, one cutting the bottom of the chest, and one in the heart. It will be a long time before we try and get another! This is a plain unvarnished account of an incident which must, I think, be unique in the annals of tiger shooting. At least, I have never heard of a lady being hurled out of a high machan by a climbing tiger and her husband killing it up in the air with his last cartridge."

The tiger as a man-eater varies clearly in several areas of technique and habit compared with both the lion, his largest rival in size, and the leopard, the differences concerning the latter cat and the tiger covered, at least from the basis of Jim Corbett's viewpoint, in the leopard chapter of this book. I know I promised not to bring Corbett back into this, so hope I may be forgiven for a couple of purely peripheral observations simply because that individual did his man-eater hunting under conditions that no longer exist and are thus interesting as a point of comparison. Yet, before we gits over our 'ips, it's best we have a look at the tiger on his own before comparing him with any relatives.

I suspect, without running tallies on my pocket calculator, that with the exception of the leopard, the tiger was, until very recent times (as close as 50 years) the most widely distributed of the great cats. Old African hands, I can assure you, have never had to do battle with the widely envisioned "lions and tigers" as perceived by the general public. True, the lion was within the past 200 years widely distributed in Asia and even India, but now exists in that subcontinent only in the Gir Forest where it is distinguished by little more than a slightly broader muzzle than the standard East African issue or is perhaps identical, dependent upon whose taxomony one chooses as gospel. If you think the tiger has problems, the Asian lion must have originated the principle of diminishing prospects. Things don't look good for the few left, heavily protected or not.

I was, however, reasonably surprised upon reading in the late paleoanthropologist Louis S. B. Leakey's 1969 book, *ANIMALS OF EAST AFRICA*, a part of the National Geographic Society "Wild Realm" series, that there probably *were* tigers in East Africa based upon a fossil jaw recovered from Bed II in 1957, from the now famous Olduvai Gorge.

This largely contradicts—or appears to—the idea that since tigers do very nicely in freezing weather and obviously suffer from the heat of much of their current range, often lying in the water or very deep shade during the heat of the day, that they were originally predators of the far north, gradually filtering south.

Putting the tiger and the lion in the same ball-park points up some interesting comparisons. Perhaps it's not the most logical first consideration, but lions are still doing a pretty good business because there didn't happen to be any Maharajas in Africa, although the British equivalent of U.S. $150 worth of Denis D. Lyell's improbably titled *AFRICAN ADVENTURE* (John Murray, London, First Edition, 1935) points out that few men since, for an arbitrary date, 1890, and the date of publication (when lions were considered vermin over much of their territory) ever shot many more than 50 or so, despite their fame as hunters. A couple are guessed by presumably unbiased contemporaries at around 200, but I strongly suspect this a bit high for a guesstimate. One very highly respected lion hunter opined that it would be impossible to hunt and kill on foot 100 lions and stand a chance of surviving. I'm inclined to agree. They sho' does bite!

Further, the history of Africa is not built around the sociality of abundant labor such as India, steeped in the pigeon-holing that the caste system could provide. There were no real beats, hunting elephants, drivers and the rest of the *pukka* rot that brought down the tiger. Had there been, the lion would be nothing but a remote memory. He's a social animal, living lazily and generally ineffectively in groups still called "prides."

I remember very well an incident told me by the late Peter Hankin who was himself a victim of a man-eating lioness. With one client and a couple of his staff, he was tracking what looked to be a very big male lion. It showed up at close range on a rocky, heavily vegetated hill, apparently on cue for the client's bullet. Peter, however, gifted with what I always think of as prudent peripheral vision, urgently suggested that the client not shoot. When asked to desist at the rates Peter was charging, *bwana* Hankin pointed out, I believe, 22 *other* lions who were taking a rather intense interest in the proceedings. That, ladies and gentlemen and members of the press, is a lot of lions. Mr. Hankin and his paying guest did an excellent job of respooring to the Land Rover and returned to camp where I presume they had the sense to drink something suitably fortifying.

I once walked good-morning-madam neat into six most irritated lions, none of which showed any indication of a need for dentures, and I will not forget it quickly. Another night, in Ethiopia, I had a running, scoreless skirmish with at least 11 visible lions. The odd thing about large collections of well-maned lions is the uncanny way they have of turning up immediately after your client has shot one that looks as if it just left a Hari Krishna barber shop. While your lion looks like the before picture in an ad for mange cure, the assembled pride all sport manes that would cause them to be instantly hired by MGM or the Where-To-Go section of a sporting magazine. Tigers, on the other hand, are notably unsocial and rarely embarrass Indian *shikaris*.

It appears high time that we got into the tiger as a processor of people meat rather than trying to pass his chapter off as *The Child's Guide To Garden Carnivora*. To be perfectly honest, I don't know what I'd do without "Table Two" of Peter Turnbull-Kemp's "Age and Condition Data on 241 Known Man-eaters at Death" to be found in *THE LEOPARD* (Howard Timmins, Cape Town, 1967) which I have no knowledge of having been reproduced elsewhere than South Africa. Turnbull-Kemp is a great writer who, handily enough, writes about my favorite animal, the leopard, in terms that only a professional game ranger or hunter could. I repeat that I know not if the volume has been reproduced for sale in other countries, but if you want my copy, acquired under duress in Botswana, you had better come armed. . . .

The point of this exercise is that the tiger is definitely of a different man-eating temperament (nonetheless deadly, though) and of apparently different collective motive than the lion or the leopard.

On a practical basis, let me demonstrate, courtesy of *Nkosi* Peter and his superb table, what I mean.

The data is based upon what Turnbull-Kemp considers reliable information, which is good enough for me. In his short introduction in this portion of the book (Chapter Ten) before the Table, he points out the obvious differences between lions, leopards and tigers as man-eaters on many bases, although both tigers and leopards are correctly considered "solitary" insomuch as they do not display group behavior as do lions.

Based upon a breakdown of confirmed man-eaters, a sampling of which was composed of 89 lions, 74 tigers, and 78 leopards, it's difficult to find a statistical variation in validity of what I seem to remember as "reliable sampling." Perhaps it would be literary piracy, but considering the completely statistical nature of the table, by Turnbull-Kemp's own commentary assembled from other sources, I would like to extract the obvious differences in categorical behavior from lions, leopards and tigers.

The first screaming difference in percentages expressed as statistics is in the initial category, concerning "age." Of the tigers, age, as subdivided by Uninjured / Injured by Man / Other Factors (presumably porcupine quills, etc.) / and Teeth Affected by Age; the tiger glows as if radioactive as being the most reticent of the man-eaters. The totals for the previous groupings for lions was a mere 18 percent. For leopard, the great hunter, it was even down to 11.5 percent. But, for the tiger, it was a dizzy 55.1 percent! Conclusion? Sick, injured, old tigers eat people on a percentage basis far more frequently than do the nice, healthy sleek tigers.

Interestingly, it was the leopards that clearly took the lead among "Mature" man-eaters, most especially those who were "Uninjured." Under the heading "Mature, Uninjured," the leopard swept the contest with an 84.7 percent correlation, whereas the tiger rated only 47.3 percent and the lion but 32.6 percent. This is a critical statistic because it tends to bear out the concept which I, at least, champion, that the leopard is a natural, matter-of-fact, everyday man-eater of primates, be they baboons, gorilla young, monkeys, chimps or people.

(128)

The "Immature" category (three years or under) goes to the lion over the tiger by nearly triple; 49.4 percent against 17.6 percent. The leopard hardly counts, with only two immature, uninjured cats involved out of 78, only 2.5 percent. One of the obvious reasons for this is the same as why we don't tend to have waves of man-eating from Siamese cats: they're too small. A leopard at less than three years would not be inclined to take on game as big as a man.

After having waded barefoot through all those fang-studded statistics, I suppose it's time to draw some conclusions. The first, and most pertinent to this tiger chapter, is that the tiger, once he's had some practice, is pretty fair at his man-eating trade, but he's not as inclined to become involved as is the lion or leopard. It would appear, contrary to the reports of African hunters on man-eaters contemporary with Corbett's Asian experience, that the tiger is statistically different from the lion and leopard, at least on the basis of Turnbull-Kemp's *Table Two*. To justify the conclusion reached by Corbett that only tigers incapacitated through injury (infected porcupine quills, gunshot wounds, etc.) ever turn to man-eating were his own experiences. Every man-eating cat he killed was or had been partially crippled. As we've seen, this is by no means the case with the lion or leopard, and seems to be less the case with the modern tiger. But, at least at the turn of the century, man-eating tigers appear to have been generally injured.

One point covered by *Table Two* does point up a similarity to some degree between lions and tigers. Although the lion had a slightly larger sampling, and the animals studied were killed much more recently than Corbett's dozen or so tigers, both species were within about 5 percent of each other in terms of condition at death. Of 74 tigers, 59 or about 75 percent were in "good" condition. In the case of the lions, 67 out of 89 were ranked the same, which is about 80 percent. It's interesting to note, however, that almost 94 percent of the 78 leopards were in fine fittle. But, then, why not? People are notoriously nutritious.

If your interest runs to the man-eaters of old, there are reams of books from which to choose, although they might these days be considered almost capital investments! Corbett is the primer and most libraries carry at least some of his work, and much more is available through secondhand booksellers. On the reasoning that if I advise the purchase of, for example, the Corbett omnibus, *MAN-EATERS OF INDIA* (Oxford Press, New York, 1957), I should also vent my extremely low-profiled opinion that one is better without Kenneth Anderson, who, in my personal consideration, is a bad imitation of Corbett. He closely followed Colonel Jim, literarily, and wrote to formula such blood-curdling chapters as entitled by the stirring sobriquets of "The Striped Terror of Chamala Valley" or "The Spotted Devil of Gummlapur," "The Mauler of Rajnagara" and even the "Marauder of Kempekari." There is a name for much of this material which, according to a friend of mine, certainly not me, is synonymous with that matter which issues from the south end of a bull heading north. Unfortunately, I tend to see Mr. Anderson quoted in truly erudite works of large *carnivora* being as gospel as Baby Jesus and Sunday morning. If you should doubt my observations, note that of Mr. Anderson's three books of which I am aware, *THE BLACK PANTHER OF SIVANIPALLI, MAN-EATERS AND JUNGLE KILLERS* and his first, *NINE MAN-EATERS AND ONE ROGUE* (the rogue presumably thrown in in case the reader tired of man-eaters) all are by different publishers, chronologically E. P. Dutton, New York, 1955; Thomas Nelson and Sons, New York, 1957; and finally George Allen & Unwin, Ltd., 1959; the American edition by Rand McNally in 1961.

To be completely frank, for every hour I spend at what Ruark called the "Iron Maiden" pounding out material, I spend at least five in research, digging out reliable tales such as the one you just read concerning Mr. and Mrs. E. A. Smythies. Do you have any idea what is involved in getting hold of a copy of the January 30, 1926 issue of the Bombay Natural History Society's *PIONEER*? Ha! And you think hardback books are getting expensive? That's why I am not about to give away any more source material on early people eaten by tigers. Anyway, you might not buy my next book if I did!

Today is another matter. I know, whaddya mean today! You thought I said that the tiger has been closed down as a species, eliminated as a menace, terminated as a threat to man. Nope. They still bite.

If your sentiments toward endangered species *really* run deep, I mean past the button-wearing stage, may I gently suggest that you visit that gem of tropic charm, the emerald necklace of the Bay of Bengal where the *Ganga Mai,* the Gangese, and the Brahmaputra Rivers enter the sea, that charming little corner of torrid hell known as the Sundarbans. Who knows, even today you might give your all—literally—to an endangered species.

Okay, let's not lose our perspective: you are not likely to be eaten by one or more tigers even if you happen to be an Indian woodcutter, although if such is the case the odds have vastly improved. Still, the really "in" place for getting eaten by *Panthera tigris* is the Sundarbans, which is sort of odd because they're not particularly attractive to man nor beasts, being mostly mangrove swamp islands or, as it is pronounced where I live at the edge of the Florida Everglades, *"Swowmp."*

I have a suspicion that any place the Gangese and Brahmaputra get together could not be much more unsanitary than ebb tide at Coney Island. Well, whatever the reason, the region appeals to tigers to beat hell. Largely mangrove, which proves that tigers are not guilty of either forethought or judgment, the Sundarbans still have the reputation of being the most likely place to get tigered that comes immediately to mind beyond having a pass key at the Bronx Zoo. In fact, the whole area has been famous since the first European exposure as being notorious for man-eaters, reports going back to the mid-1600s of tigers actually swimming out to sea, climbing aboard boats and eating folks. In fact, these reports are so strung-out and consistent that their veracity can't really be reasonably doubted. According to Ricciuti, presumably quoting another source, 275 people were taken by tigers between 1961 and 1971 in the Sundarbans and at least five lives were taken during the month of April, 1973, when the local people were hunting honey.

During the recent Vietnamese conflict, there were several cases of genuine tiger attack on U.N. Troops, including Americans. In fact, one of the most realistic portrayals of the stealth and surprise factor of a big cat like a tiger was in the hit movie, *APOCALYPSE NOW,* which you probably saw, and which had an excellent tiger scene. Unquestionably, soldiers from both sides were killed and eaten by tigers—as well as by leopards—and much of this may have been generated by the abandonment of bodies in the jungle when fire-fights or other conditions made body retrieval impractical. The tiger is no less a scavenger than the lion.

It has been postulated that, over the last 400 years, at least a half-million Indians alone have been eaten by tigers and, considering the vastness of Asia, the totals of humans must run several times that number. Today, it is common to think of the "Bengal" tiger as being the predominant species or subspecies, however, there are other branches of the family well worth noting. The Siberian snow tiger is the largest of the clan and, although it's stock is sinking like the rest, is certainly an impressive creature by any standard.

There is a reliable tradition that, in Russia, in the *taiga,* that marvelously mysterious veil of evergreens that so fascinated the late zoologist, Ivan Sanderson, to the extent that he believed all sorts of presumed extinct animals might live there, the tiger yet holds a traditional sway if not physical. It would seem that it was a custom of the Cossacks to lash criminals to certain trees as punishment of a one-way variety: tigers were familiar with the trees and ate the offenders, probably developing regrettable manners in the meanwhile.

The tiger, as a man-eater, although astonishingly effective, is probably the least *innately* offensive of the great cats that are established as eaters of men, the truth lying more on the side of the historical traditionalist inasmuch as the majority do seem to prefer nonhuman fare. Unfortunately, the tiger turned man-eater is very possibly the most dangerous animal on earth. If you have any doubt, see how many survivors there were of the Champawat tigress.

I think you get my point.

CHAPTER SEVEN

MAN ON
THE MENU

As you have probably noticed, this cheery little book offers a wonderful assortment of creatures great and small (*etc. ad nauseum*) who share a varying talent for eating people. Some—sharks, lions, tigers and such— are a bit better known for their endeavors in this area, but the work wouldn't really be complete without a mention of several of what are surely the more exotic of this somewhat specialized grouping, who have the rather chilling habit of looking at the Spitting Image of the Lord of Creation and seeing only a lump of bipedal protein. Among the most deliciously fascinating and genuinely terrifying is a beautiful little red and silver messenger of death common throughout much of tropical South America. In his most effective form, you'll find him listed as *Serrasalmus nattereri,* one of the most common of the more than 20 forms he takes. But, in Brazil, Guyana or Venezuela, just make a big splash in the nearest stream and yell "Piranha!" if you want to gain some rapid local attention.

Of all people, it was Teddy Roosevelt who probably did as much to advance the negative side of what he called the "cannibal fish" as any other writer. (This was a correct nomenclatural assessment: a whole milk can of piranhas traditionally reaches its destination with one surviving inhabitant suffering from acute indigestion. This artificially, if effectively, inflates the cost of piranhas where legally bought and sold, which does *not* include the United States. Roosevelt, in his *THROUGH*

THE BRAZILIAN JUNGLES (G. P. Putnam's Sons, New York 1914) undoubtedly was as impressed by his first demonstration of the enthusiasm of a school of feeding piranha as was I. Teddy called them "the most ferocious fish in the world," and I have strong inclinations to agree with him. Again, Roosevelt was careful with his words. "Ferocious" is not, by my interpretation anyway, another of those adjectives that ascribe human values to animals such as "crafty" or "vicious." It's from the Latin *ferox,* meaning "wild." If you ever see a school of piranha get warmed up to some free lunch, I very much doubt that you'll pick any bones (excuse the expression) over the use of the description, "ferocious." The experience will scare the bloody hell out of you. Heavens! I have one ex-jaguar hunting client who still breaks out in hives if he sees a glass of water, and that was from a demonstration I put on for him 17 years ago! Of course you will appreciate that, as a hunter, I never, ever, exaggerate. . . .

Since I have largely composed this manuscript of the experiences of others, I will take this opportunity to give a personal description of my first exposure to piranhas in the wild. I was a guest of Count Andre Rakowitsch, a fascinating character who, although a titled White Russian, was a decorated panzer commander for the Germans, whom he preferred, if only slightly, to the Reds of his own country. Andre had managed to emigrate to Brazil after the war, avoiding an all-expense extended vacation in lovely, cool Siberia. As a surveyor in the interior, he befriended several of the local tribes, particularly in the area of Bananal Island, the largest river island in the world, in the region of the Araguaia and Tapirapé Rivers. After many years, he flew in the component parts for a houseboat to act as a floating hotel and mother ship for a fleet of smaller skiffs to take advantage of the phenomenal fishing in these rivers. I was aboard his boat, the *Brazil Safari* in the mid-1960's to sample this and to act as a representative for him through a travel agency I owned back in New York. Anxious to have some action shots for the photo slide presentation I had in mind, I asked him to show me some piranhas. This, for sure, he did.

We weighed anchor and ran up the Araguaia to its juncture with the smaller Tapirapé, which Andre advised was a spot of particular concentrations of these killers. Taking the skinned carcass of a Caiman crocodile about two feet long, he wired it to a stick by one end of the trace and the croc's snout by the other. We walked up to the bow of the moored boat and peered into the murky, softly swirling water. It certainly *looked* peaceful.

Over lunch, I had been picking Andre's brains about the piranha generally and learned that he believed that commotion in the water was far more important to attract them—at least this "red" species—than was the presence of blood. As a demonstration, he let the bloody carcass soak quietly for several minutes with no reaction at all from the piranha presumably below. Then, taking the branch or stick, he began to slap the carcass against the surface. Not fifteen seconds passed before there was the first flash of movement, a burnished gleam of silver. Then, things started to get out of hand.

For many years I have tried to dream up a really adequate description for a close-up view of mass-feeding by red piranha. I will have to stick by my original contention that it has the appearance on the surface of the water boiling as if in a large, glass crystal bowl while red and white flashes of what look like poker chips the size of a big man's hand swirl madly about. The sound of the jaws, at least from close range, is truly chilling although unmatched by any sound I have ever heard.

Although I don't believe that the tales of whole cows or people being devoured down to a skeleton in a matter of 30 seconds are accurate, they're not terribly far off the mark. In a big school of several thousand *Serrasalmus nattereri,* which I believe is not rare, a 150-pound man could probably, in all practicality, be eaten in about five minutes, possibly a bit less.

I spent most of the next afternoon photographing feeding piranha from a leaning palm tree over the river bank, working close with a 200 mm lens on my Pentax SLR. You can make book on the fact that I did nothing fancy. There was absolutely no question in my mind that anything edible, including me, would not live longer than about one good scream if it dropped into that water.

Considering that the red piranha is only about eight to eleven inches long in a good-sized individual, the sheer destructive power of the creature is purely astonishing. A large or medium school would be sure death from shock and loss of blood, even in case of an almost instantaneous rescue, which would be highly unlikely. The speed of feeding is much faster than a garbage disposal and the jaws of each individual, a dried set of which I am holding in my left hand right now (and have just used to slice away a paper-thin slab of callus from the base of my left ring finger), have teeth as sharp as a paring knife. I have seen several—two or three—people who bear piranha scars on their legs, and one old man on *Ilha do Marajo,* the Amazon delta, who was missing all the toes of one foot and quite a bit of healed flesh as well. He was hit in extremely shallow water at a ford and was able to leap to safety. He told me most of the bites happened out of the water where they were completed by piranha who were still hanging on as he sprang free. That had happened about 1950. The one other man I spoke with through an interpreter was an Indian who was missing just two scoops of meat from one leg, leaving a characteristic roundish, hollow scar rather like one would have if attacked with a large melon ball cutter. He advised me he did not even know he had been bitten until he left the water, so sharp were the teeth the bites were painless.

Literature is, of course, full of bloodcurdling tales of the piranha and there is no doubt that many men have been killed by these scourges. Sasha Siemel's arch enemy, Ricardo Favelle, was so terribly mutilated by piranha that he shot himself. Eduardo Barros Prado reports (fancifully?) a girl who purposely went swimming in piranha infested waters of the Canuma River of Brazil and was slaughtered, only her skeleton, swim suit and red hair left after she dived off her uncle's yacht. Sorry, this would be the result of running afoul a school of piranha, but the red hair touch just smells a bit Hollywoody for me. Colonel Fawcett, who had his expedition wiped out by the Xavantes Indians quite near where I was photographing piranha, records among his juicier incidents, a Brazilian trooper who fell out of a dugout on the Corumba River, which is a particularly bad one for *Serrasalmus nattereri,* who was recovered dead, his lifeless fingers supporting from the gunwale only the top half of his body, the legs and waist having been eaten away by piranha.

I have caught hundreds of piranha on rod and reel, often several on a single cast when using a multiple-hooked plug. In fact, they are a pain in the neck when trying to catch peacock bass and other sporting species of these rivers. Besides the "red," the areas I know best also have the "black" piranha, which is slightly bigger and much darker. I was told that it is not feared and eats only fish. The biggest of the piranhas I have seen was from the Xingu and was phonetically called the "Shipita," although this may well be the "Xipita" as in Brazilian Portuguese, "X" is pronounced "Sh." Andre told me, according to my now faded notes, that it would reach 14 pounds in the Xingu basin. Fortunately, they are not a large school fish and do not have a man-eating reputation, which is just as well! About six of those boys could clear out a joint session of Congress!

One of the most terrifying aspects of the red piranha is the fact that it was until fairly recently imported into America in somewhat large numbers for aquariums, probably sold to the same fringe element that keep king cobras and African lions. The big problem is that several have been caught wild in Florida (already overrun with walking catfish, armadillos from Mexico, African giant frogs and Christ only knows what) in areas around Lake Okeechobee and elsewhere. If the red piranha ever gains a foothold in temperate America because some nitwit purposely released them into suitable waters, we're going to have a natural catastrophe that will make the killer bees and fire ants very secondary problems in comparison. If it will cheer you, several ponds in which piranha have been caught have already been completely poisoned, killing every fish in them to be sure no piranha survived. What if those ponds had been large canal systems? Or what if piranha are already in them?

*   *   *

As long as we're on the little stuff, a word should be given the social insects, especially the ants. Bees or fire ants will sting you to death and termites will collapse your house around your ears, but only the African driver ants will actually eat you, which is an interesting distinction, if you happen to live in man-eating ant country.

It may very well be that the driver ant of Africa, a very robust example of the family supported by the same incredible organizational talent of other cooperative insects, should be included in the category of "questionable" as a man-killer and eater, but if you have ever witnessed an attack by a column of these terrors, you will probably go along with my presumption of probability that some helpless babies, laid aside while the mother works in the fields or some wounded and incapacitated adults, ill or victims of animals or war have, indeed, been eaten by the awesome driver ants.

The only place I have ever seen this species was in Sidamo Province of southern Ethiopia in January of 1968, camped along the Dawa River, quite near the Northern Frontier District of Kenya. I am sure they exist elsewhere, but not seemingly in the more familiar Zambia, Botswana or Rhodesia of most of my professional hunting career. On this safari, the son of a professional hunter with us, along for a brief holiday, had brought his girl friend, an American Peace Corps volunteer from Florida.

We had a particularly pretty camp, set in a clump of trees handily spaced for shade in daytime but not cramped. The sun had been down about three hours, making the time a bit after nine and we were sitting in camp chairs at varying distances from the fire. Everybody had a nightcap and the conversation was pleasant in the semidarkness, distant jackals warming up far away. I was about to say something to somebody when, like a siren starting to build, a wild scream built into a towering shriek and the girl friend threw herself into a maneuver it might have taken Nadia Comaneci months to master. Stunned, everybody thought she had completely blown a main terminal as she tore at her clothes, her skirt, underwear and what all inscribing ghostly arcs through the half light.

The hunter's son, who was nearest her, leaped up and started the same performance, cursing and slapping at his legs crazily. "Siafu," he kept screaming, which, when it sank in, galvanized his father and me into movement. As we dragged them nearer the fire, so we could see, it was clear that their legs and lower bodies, particularly the girl's, were almost swarming with big, black *siafu* or driver ants. I had read of their scary habit of sneaking up the legs of someone in their path, somehow unfelt, then biting on some signal given by their leader, all at once. Apparently, it was true, in spades!

Believe me, brother, nobody was in the slightest embarrassed about the removal of those ants from some traditionally delicate regions of the assembled anatomy. Both the boy and girl were in agony until we could scrape off the biting, stubborn and presumably at least slightly toxic creatures. I used a butter knife, often having to return to get the heads, which kept hanging on after the bodies had been torn off. Keeping them well away from their chairs where they had originally been swarmed,

(137)

a good coating of antiseptic and analgesic creme under some fresh clothes helped a lot. But for some nasty marks the next morning, there was no permanent harm. Just how lucky they had been, I couldn't tell you, but it would have taken a lot more ants than were on them to have done genuinely serious harm although the pain was severe. Still, a small child might well have been killed and, judging by the size of the column, eaten.

With flashlights, we watched the army, about 10 ants wide and by the time we saw it, perhaps 100 yards long, disappearing under a flat rock about 30 yards away. I have since learned that these *siafu* bands may be several thousands of yards long and chase whole villages away. Generally, the Africans are delighted as the ants eat all the vermin left behind. That big game lives in constant fear of these constantly migrating rippers is doubtful, but they have been reliably recorded as having killed and eaten dogs, pythons and at least one small crocodile (a pet) which was not eaten. I suspect *siafu* are fairly rare. That's just as well. . . .

The ant species that fictionally opposed Charlton Heston in the film version of *Leiningen and the Ants,* the tale of the army of ants who attacked the plantation somewhere in Amazonia and were defeated by the planter, apparently do not even exist! *Marabunda,* as they were called in the story are not among the many South American ants, which do include some harmless but entertainingly talented species. Sorry about that, movie fans.

*   *   *

I'll bet you thought we'd never get to it; that darling of all the real horrors, the true lurking culprit (would you believe dangling?) that hungers for the flesh of explorers—particularly those very few who can read and write—cramming down whole men with the same inexorable efficiency of a troop of cub scouts at a fast-food hamburger joint. Yes, folks, its time for the GIANT SNAKES! Yessir, you've come to the right place. We've got pythons that crush, anacondas that strangle men like rotten bananas and even, at slight extra cost, boas that constrict like the Federal Trade Commission run amok!

I mean, they do, uh, don't they?

Take heart, for, verily, it appears that sometimes they *do* kill and eat people. Perhaps you share my craving for a great sigh of relief. At least one more semi-illusion preserved.

You will be relieved to know that, in September of 1965, the Reuters news agency reported from Rangoon, Burma, that an eight-year-old boy was swallowed by a python in a village called Ye. "Villagers killed the 15 foot python and took out the boy,"—obviously somebody on the spot was thinking quickly—"but he was dead. His bones had been crushed."

I hardly find it worth a paragraph to reiterate what is by this date fairly common knowledge, i.e., that constrictor snakes do not "crush", "crunch" or otherwise obviously damage their victims, beyond bad bites, merely literally constricting the rib cage from expanding so the victim may not draw in fresh air and thus dies of asphyxiation.

That information, while reasonably accurate, might not have been terribly comforting to an African working at a mine in South Africa's Transvaal in 1961 who—in a most unAfrican manner, as most are scared to death of snakes—grabbed a python he saw in a nearby patch of bush by the tail. It turned out to be a bit more python than he'd planned on, as it promptly threw two coils over the man's chest, bringing him to the ground. After a pretty fair fight, the miner got loose (or, perhaps the snake did) and went back to work. The following day, he complained of severe head pains and was sent to the hospital. The next day he was stone dead. Doctors found he had a ruptured spleen and damaged kidneys! I suppose this proves that although a constrictor snake doesn't crush his food, he *can* keep a pretty tight hold.

There is a persistent and seemingly well-documented case of a 14-year-old boy having been killed and eaten by a python in the East Indian Taland Island Group. I've found four serious references to it and it appears genuine, but such happenings are very rare.

The reason that the big snakes are so rarely fatal to man is two-fold. First, man is generally too big a prey to swallow, although some humans are unquestionably killed and abandoned after running afoul of one or another of the big boys and, secondly, who in their right mind fools around with a really big snake in the wild without knowing what they're doing?

Except to admit that giant snakes *do* rarely kill people and even more rarely eat them, I suggest that the matter be left here as overly difficult to document on an extended basis.

*   *   *

From a basic posture of cringing self-defense, there are three species I would like to mention in passing, for the reason that I don't want a critic to get the idea that I was unaware that they eat people or portions thereof.

The first is the Komodo dragon, a giant form of monitor lizard that gets to about 10 feet and is tough enough to regularly kill and eat deer, wild boar and at least one person, according to the book *GIANT REPTILES* by Sherman A. Minton, Jr. and Madge Rutherford Minton (Charles Scribner's, New York, 1973). It is confined (thank Heaven) to a very small range of islands off Asia, where it is the dominant life form and predator.

The second man-eater, I am unable to document, but recall reading what I took to be at least one multiple eye-witness account of an American serviceman, during World War II, somewhere in the Pacific, being swarmed, killed and eaten by the

Black Triggerfish (*Odonus niger*). My memory may be faulty as to precise species, but I am sure it was some black form of common coral reef fish.

The third animal that very commonly kills and eats man may be lying at your feet right now, the domestic dog. I have been twice bitten by dogs (only partly eaten, an ear torn off, or mostly so), once by a chow and once by a poodle. My step-grandson is only two years old, but he's been bitten, minorly, in the face, three times. When my first cousin, Mrs. Chandler Pitcher, was living in New Jersey, her next door neighbor was killed, dismembered and eaten by her pair of pet Dobermans. The documentation of pet eating owner, forgetting the severe bites, would fill quite an impressive space, a space I won't try to duplicate here because, as a domestic animal, it doesn't really belong in the zone of knowledge of this book as I perceive it. That doesn't mean I dislike dogs; quite the contrary. I am careful around them, though, particularly with strange ones, but try never to show sheer, stark terror from the neighbor's Schnauser. After all, I believe in tough pets: I keep the most savage, attack-trained goldfish on the Eastern Seaboard. My hamster has a black belt in Kung-fu.

If you think you have problems, consider that people, especially children, have been killed and eaten by domesticated barnyard variety pigs. Apparently, the old line about "not having so much fun since a hog ate my little brother" may have a nasty grounding in fact. It may not do much for your appreciation of bacon, but most front-line soldiers will confirm that pigs are also purely fond of eating bodies. One hears a lot of horror movie tales about dungeon rats killing and eating people. I'll go along with the eating part, but a swarm of rats killing a person is tough to prove, except through rabies or bubonic plague and that doesn't count.

Vampire bats regularly drink the blood of sleeping humans but only very rarely has a death through blood loss been reported.

Oh, yes, I forgot, during the 1940's in Zurich, Switzerland, an Indian elephant in the municipal zoo killed and *ate* a zoo employee named Bertha Walt.

* * *

All you would have to do to figure out that I'm not very good at predicting the future would be to check with my stockbroker. There is one area, though, in which I'm willing, as it were, to stick out your neck as well as mine: I predict that death by and attack by alligators in the American South will become more and more common the more closely the alligator is protected and the more habitats of man and 'gator overlap.

I was rather surprised that Roger A. Caras, the well-known television animal commentator and author of, among other works, *DANGEROUS TO MAN*, which he revised in 1975, made the comment in that edition that, "As with the giant Gavial (*Gavialis gangeticus*) of India, there are no authentic records of man-eating by American alligators." Mr. Caras and I do not always see eye-to-eye, which is understandable considering that I am a professional African hunter and he the author of such statements as he spontaneously offered the audience of ABC-TV's 20/20, speaking with Hugh Downs on a segment in January of 1981 about endangered species:

"He who destroys man's creation is called a vandal; he who destroys God's creation is called a sportsman," intoned Mr. Caras righteously. He was about as correct on that point as he was concerning the alligator.

The following is an Associated Press report taken from the Naples (Florida) *DAILY NEWS* for Sunday, August 19, 1973 under the headline

GATOR IS SUSPECT IN GIRL'S DEATH

SARASOTA, Fla. (AP)—"I grabbed her hair but it kept pulling me down too," said the father of a 16-year-old girl who was partially devoured after apparently being drowned by an alligator.

A state wildlife official said if an autopsy confirms that the alligator killed the girl, it would be the first documented case in the history of a state known for its number of the reptiles.

Bert Russell Holmes said he and his daughter, Sharon Elaine, were swimming in a lake at Oscar Sherer State Park south of Sarasota on Florida's lower Gulf Coast during a family outing Thursday.

He said his daughter was in the middle of the five-acre lake and he was standing in a shallow water when she suddenly yelled for help.

"She screamed 'Daddy' like something had got a hold of her," said Holmes. "I turned around and looked but Sharon wasn't even in sight. That girl didn't even clear the top of the water with her head."

"Then I saw her hand come up for just a moment," Holmes said. He said he swam to where he had seen the hand flash on the surface but couldn't locate his daughter.

"Can you see her?," he said he shouted back to his wife, who was standing on the bank.

"She's right behind you," she yelled back.

He turned, spotted a tip of her finger as it submerged and

then dove after her, managed to grab her hip-length blond hair but couldn't pull her to the surface.

"I dove again and again," said Holmes.

Several hours later, after divers had searched the lake, the mauled body of Miss Holmes was found on the lake's bank. It was guarded by a 10-foot alligator which threatened the first officer to approach but crawled into the water as others gathered, police said.

"We killed the alligator Friday morning," said deputy Eddie Palmer, "and found a hand and an elbow in its stomach."

The Sarasota County medical examiner said autopsy reports would be released Monday.

Colonel Brantley Goodson of the Florida Game and Fresh Water Commission said that although there have been scattered reports over the years of alligators attacking humans, there has never been a recorded case of one killing a person.

"This is the first time we've even heard of an alligator attacking and possibly killing someone," he said. Sources in Louisiana and southern Mississippi, also populated with alligators, could not immediately recall any recorded deaths by the reptiles.

Officials said swimming had been banned temporarily two weeks ago after the alligator was first spotted. But a cautionary sign was removed after several days passed without an additional sighting.

Holmes, whose occupation was not immediately determined, said he intends to sue the state for failing to properly warn swimmers.

He said the only sign at the lake read: "No Lifeguard on duty—swim at your own risk."

The following week, the death was confirmed, as if that was necessary considering the 'gator was guarding the body and contained a "hand and an elbow." There are, however, in the follow-up Associated Press article, some interesting comments from a research biologist on the subject:

"People monkey around with alligators, they feed them, take them as pets . . . It's just risky to tame a large dangerous animal," he says.

Sound familiar?

"There's not too much a person can do in this case," says Lovett Williams, a research biologist with the Florida Game and Fresh Water Commission in Gainesville. "With an alligator this size the biggest man that ever visited that park would have been in the same situation."

"If an alligator of 11 feet attacked you, he'd just crush you no matter how big you were. If he got anywhere except an arm or leg, you've had it. And if he got an arm or leg he'd twist it right off," Williams says.

But Williams says there are two things that can be done to prevent an attack like the one that killed Miss Holmes from happening again. The first admonition, Williams said, is "don't feed the gators."

"Alligators have a natural fear of man, but it's overcome if you give them handouts. This alligator in a sense became domesticated and had no fear of people."

But officials say rounding up the alligators has been a problem. First, there are more of the scaly reptiles than ever before. Due to legislation limiting the use of alligator hides for billfolds, belts, and purses, officials estimate that Florida's alligator population has doubled to 300,000 in the past five years.

Furthermore, officials say they are overwhelmed with requests to remove the creatures from lakes and ponds whether or not they pose a threat to human life.

The alligator is unquestionably suffering from the energy shortage as much as everybody else as he was reported as recently as 1791 by Quaker traveler William Bartram as having clouds of smoke issuing from his dilated nostrils. I guess OPEC is getting to us all. . . .

Although poor Sharon Elaine Holmes is the only fatality of a 'gator man-eater to date, there have been many other attacks that logically would have led to the animal feeding on his kill, so far as I can conclude.

A good example took place on September 16, 1952, when ten-year-old Parker E. Spratt went fishing with a neighborhood Coral Gables girl, Jerry Gustafson, in a rock pit near their houses. Before either knew what had happened, a seven-foot 'gator burst out of the dark water, knocked down the boy and closed its jaws over little Jerry's arm, dragging her under.

For some reason, although the girl was half-conscious, the reptile rose up again and somehow lost his tooth-hold as he tried what was the standard maneuver used by crocs and gators to tear off limbs, i.e., to roll with it. Seeing his chance, young Parker, holding on to a tree root over the pool, pulled his friend out and got between the 'gator and the girl. Gradually, he half-climbed and crawled with the girl ahead of him to safety as the 'gator waited below.

He was personally decorated by President Truman with the Young America Bravery Medal for his rescue. Jerry's arm was broken and badly lacerated.

Now, if that cub scout hadn't been there to save the nine-year-old would you believe that she would not have been killed and eaten?

If she had not been rescued, she would probably fallen into the category developed by the professionally skeptical which brackets the death of another nine-year-old, a boy who was found mauled by alligators in 1957 in Eau Gallie, Brevard County, Florida. "Conservationists" maintained that he had drowned before the 'gators got him, but I can find no record of this substantiated by an autopsy. When almost the same thing happened again in 1959 near Daytona Beach, presumably the conservationists were again handy to preserve the alligators' civil rights. Several years before, a woman had survived an attack in the Weekiewatchee River on Florida's West Coast with only 23 stitches in her arm and hand when a big gator apparently changed his mind in mid-attack and withdrew.

Just as it seems impossible to convince visitors to national parks that bears are dangerous and wild, so it is with the 'gator. I must personally know a dozen otherwise rational people who carry—of all bloody things—marshmallows to feed the gators at most of the twenty-odd golf course water hole complexes just here in the small City of Naples, Florida. Presumably, it won't be long before they start putting their grandchildren on the 'gators backs and then will wonder why things suddenly turned deadly.

Well, it seems the old story as with other potentially dangerous wildlife forms coming into closer and closer contact with man, each exposure further eroding the natural fear of wildlife for man. I very much hope I'm wrong about my prediction, but I've got a hunch that somebody has already bet their life I'm not.

Leon Parson

CHAPTER EIGHT

QUESTIONABLE KILLERS

What is 30 feet long, a zowie two-tone black and white, eats 13 porpoises and 14 seals at a single sitting and swims absolutely any place it wants to? Wrong. It is not Dick Butkus in a checkered swim suit. It is *Orcinus orca,* who is not a has-been Roman emperor, but an adult bull grampus or killer whale, about the biggest meat-eating mammal this side of a sperm whale.

I've had a very hard few days, lately. All through the various contemplative stages of this book I had been looking forward with what was no less than true awe to researching the mayhem and blood-lust that has always tinted the waters of my imagination at the merest mention of this granddaddy of all man-eaters. Imagine! Getting paid to learn all about a genuine terror that would make a science fiction writer blush to dream him up! Think of an animal so savage that it can knock people off ice floes, flip whole Beluga whales out of the water just as it does Eskimos in their kayaks, catching them in meshing javelin-pointed teeth two inches in diameter! Dream of a brute so savage that even the *U.S. Navy Diving Manual* awards it the highest danger rating, a four-plus, and on top of that calls it "ferocious" and "ruthless;" a beast that sinks 19-ton yachts, no less! Hot damn! You want to talk about man-eaters? Now we're in the big leagues.

At least I thought we were.

It was rather the same sensation I recall upon finding out that there wasn't *really* a Santa Claus; a real end-of-the-world sensation like discovering that Hopalong Cassidy or The Lone Ranger secretly liked girls. But, after four days of growing panic, desperation creeping closer as each source material let me down, I finally had to face up to my darkest fears coming true; the killer whale was a sissy.

Well, maybe not exactly a sissy. It's pretty tough to label a biting machine as awesome as that a pantywaist, but the problem was still clear: killer whales, for whatever reason I cannot imagine, do not eat people. At least, not so that you could prove it, they don't. They have never even had the common courtesy to have unquestionably swallowed one lousy yachtsman, although they are supposed to spend most of the time they're not knocking photographers off icebergs into the water biting holes in a variety of boats and ships to get at the occupants. I can tell you, the whole project has been mighty discouraging.

Frankly, the more I have read about killer whales (imagine not being a man-eater with a name like that!) the more astonished I have been to find that there exists *no* absolutely reliable record of a killer whale eating a man. Or woman. Not even somebody's kid! Of course, with the sort of appetite the largest of the dolphins has along with its unexaggerated skill, power and rapaciousness as a hunter of warm-blooded prey including seals, porpoises, and penguins, there is bound to be no shortage of legerdermain draped over its reputation. Since it has been in captivity something more than a decade, we've learned that it's rather a pleasant and intelligent creature and, at least when well fed, does not tend to live up to its tradition of mayhem. I don't think that to date even one of those crazy trainers who stick their heads in the killer whale's mouth at marine exhibition shows has had any reason to have experienced a rather dramatic end to "ring around the collar" through loss of the neck.

Hang on now, it's not that there aren't plenty of tales which would fall into the category of what my father referred to as "sea stories" when expressing some slight question of the veracity of one of my performances in the grouse woods or duck blind. You may recall a character I remember as John D. Craig, who had a television show if memory serves, and a book by the same name, *DANGER IS MY BUSINESS* which I read to shreds longer ago than I care to recall. One of the most interesting and exciting episodes included an underwater attack by a determined grampus on a diver, who managed to keep out of the awful jaws by wedging himself into a rock niche as the killer tried to tear him free. In fact, that tale may have started my well-established, although hastily reached, conclusions about the killer whale.

It might be fair to say that killer whales *may* have rarely *tried* to eat people, one famous incident having taken place in the Antarctic during the 1911 Scott Expedition. The photographer, Herbert F. Ponting, was nearly knocked off an ice floe along with some sled dogs by seven or eight killers who even put their heads out of the water to better see him. They may have thought him a weird form of duded-up seal, which seems likely, as there hadn't been very many people to the Antarctic by 1911 in any case. If Ponting had fallen in would he have been eaten? I can't imagine why not, but, as I'm sure Mr. Ponting would agree is most fortuitous, we'll probably never know.

Many of the supposed attacks by killer whales on small boats have been attributed to this species in plain error. One famous case, in 1952 near San Francisco, was hung on a killer whale which was driven off after a bite that eventually sank a 14-foot boat with two passengers close enough to shore that they were saved. Later investigations clearly proved that the attacker was a great white shark by the dentition. In South African waters, however, a two-foot chunk of hardwood was torn out of a yacht's rudder, the attack attributed to a killer because of the teeth that were left embedded. Who identified the teeth is unclear, as is the original source of the material. I smell a rat. As a warm-blooded animal, I can't imagine why a killer would break off 10 percent of his teeth, a very painful experience, by biting a wooden rudder. Perhaps it's true, though, and he thought the rudder of the sailboat was a whale fluke.

Considering how effective a killer the grampus is, it's hard to understand why he wouldn't eat man. They are very reliably reported as tearing out the tongues of blue whales and eating away the lips. Personally, I would opine that the odd Eskimo or two has probably gotten eaten, even if only mistaken for a seal or his kayak for a surfaced, small beluga whale, one of the grampus's favorite foods. After all, an animal that eats porpoises *whole* wouldn't leave any tell-tale sign if he ate a man, anyway.

One of the odder ventures dealing with the killer whale centers around the attack by three sea creatures believed to have been killer whales on several children of Douglas Robertson, owner of the yacht *Lucette*. The 43-foot long, 19-ton schooner sank quickly after being rammed in the rough vicinity of the Galapagos Islands, although the crew survived by making a jury-rigged raft out of the wreckage. Robertson's book, *SURVIVE THE SAVAGE SEA* (Praeger, 1973) records that one of the "killer whales" if that's what they were, hit the hull hard enough to lay its head open, which doesn't sound like a very effective hunting method to me. Perhaps they just ran into the *Lucette* by accident or, when blood was in the water, did not eat the people preferring their wounded comrade, instead.

So, the meat of the nut is that the killer whale, one of the most efficient predators and carnivorous animals on earth, has never been convicted of eating man. That, of course doesn't mean that he never has, just that we can't prove it; at least not yet. As

(149)

for me, pal, when old grampus comes finning by, don't look around for this boy in the immediate vicinity. It would be just by luck to break the record.

*   *   *

The jaguar, *Panthera onca,* the great cat of the southern American wilds, is second only to the more weighty grizzly, brown, and polar bears of the U.S., Canada and Alaska as a potential man-eater. Probably because of the vast denseness of the jaguar's habitat, we know embarrassingly little about him, particularly as a killer and eater of humans.

Having been recorded by what must be considered an unimpeachable source— me—at 460 pounds from the Mato Grosso of Brazil, the *tigre,* as he's called through his Spanish and Portuguese language range looks kind of like a leopard that's been on a diet of chocolate bonbons and steroids. He once was well established as far north as the American midwest but now is left in strong pockets through the jungles south into Mexico, Central America and South America, as far as possibly northern Patagonia. Actually, he's found in all kinds of terrain where there is enough food to support a territory, a solitary over-lapping sexual arrangement a male jaguar seems to reckon ideal. Since this may include large areas of grassy cattle rangeland, there is a decided tendency for jaguars to be highly unpopular with most cattlemen's associations. The reason for including this marvelous looking hunk of teeth, claws and spots in the category of "Questionable Killers" is that as a people-purloiner, he is a decided underachiever. Easy now, calm down. Foam is hard to get off your necktie once it dries. I'm not saying that the jaguar is not a man-eater; unquestionably it is. The problem at least for the purposes of this book, is that it's really marginal when compared with lions, tigers or leopards, among which it ranks as the third biggest.

There are a fair number of parallels between jaguars, leopards and tigers; few with the lion. The first three are all solitary (except when warming up to pitch a little woo) and are territorial. They all regularly eat similar prey, according to size and availability. Yet, only the tiger, leopard and lion are regularly guilty of man-eating and I'll be damned if I can figure out why.

Taste doesn't seem a likely factor. Surely a nice, plump Indian from the Xingu or a traveling preacher from Paraguay would both be as relatively tasty as an Indian Gond, a Pakistani or perhaps a Kalahari Bushman or a Zulu maiden. Nah, that couldn't be it. Perhaps jaguars are never driven to man-eating by a shortage of natural game such as produced by African droughts? Nope. There's far less game in South America than in either Africa or most of the tiger's and leopard's Asian range.

All this may be starting to give you some idea of what we poor, selfless writers, condemned through one or another indiscretion of fine print in the production contract, have to go through to come up with material on man-eating jaguars when the species, despite looking so wonderfully fearful, hungry and fit is, in actuality, so bloody uncooperative. I'm not the only one it's left stranded, either. The late Ernest Thompson Seton, a Canadian naturalist who wrote so many animal books he must have had a deal to sell them by the pound, got caught by a very strange man-eating jaguar story which he passed along in his 1925 *LIVES OF GAME ANIMALS*. In fairness, old Ernie was almost surely duped by the quasi-official report of a reliable "expert" almost three-quarters of a century earlier, from whose material Seton drew. Here's how the tale went. . . .

The Convent of San Francisco in Santa Fe, the State of New Mexico 88 years later, probably lay under a warm and pleasant blanket of sunshine on April 10th, 1825, although we don't know the actual weather conditions when a lay brother finished his prayers and walked from the chapel, through the sacristy on his way to reach the outside garden.

He never made it. There was a growl, a scream and the peculiar silence of death.

A guard at the Convent, surprised to hear the commotion went to the sacristy to see what was going on. As he entered the dark, windowless room he, too, felt the shock of long, hooked talons and crushing, tearing teeth of the animal that would prowl the legend and literature of the American south west for the next century as the man-eating Jaguar of Rio Bravo.

As the two dead men lay on the dark floor of the sacristy, a group of monks also chanced to hear the growls and screams and one, who could not have been guilty of the deadly sin of taking pride in his smarts, also ventured into the dim chamber. Guess what.

By this time, the Convent generally was catching on that there was a somewhat irreverent jaguar in the sacristy and a party of men formed up under the leadership of the take-charge Senator Iriondo, a politician who apparently was visiting Santa Fe. As the group of rescuers ran around one end of the chamber, the jaguar apparently leaped out into the garden, grabbed another victim and hauled him back to his growing collection in the sacristy. The Senator, however, was too quick to miss his chance and had both doors of the room slammed and barred. They sent for an armed man, and the man-eating Jaguar of Rio Bravo had its career ended by a bullet through a hole cut in one of the doors.

I think it's kind of a fun story, certainly guaranteed to add rainbows of local color to the history of Santa Fe. So, it's rather a shame to have to report to you that it never happened. Ernie Seton, along with probably anybody else who ever read the semiofficial "history" of the incident had been literally hornswoggled!

James Clarke, who did the detective work exposing the fraud in his 1969 book *MAN IS THE PREY* (Stein and Day, New York) or who at least discovered the sleuth who did uncover the lie, wrote that the report on the man-killing was supplied by a Professor Spencer F. Baird, assistant secretary of the Smithsonian Institution, no less, who said that *he* got it directly from the historical records of the Convent of San Francisco. This was not true. Baird certainly convinced Seton, who stated his source as a United States Government report dating from 1857 (which must have been Baird's) but the whole incident was absolute trash. As Clarke points out, there simply never was a Convent of San Francisco at Santa Fe, there were neither lay brothers nor monks at this time in New Mexico, and even that master of feline tactics, Senator Iriondo, had never managed to get himself born. Whether Professor Baird was an elaborate practical joker—he, at least, had the courtesy to exist—or simply thought that Santa Fe would be better off with a man-eating jaguar sanctified by none less than the Smithsonian Institution, we shall never know. All we do know is that I am still having problems producing genuine, man-eating jaguars or reliable reports thereof just when I badly need them.

Actually, it's not for lack of mention of the jaguar as a man-eater that I'm reticent to give his all-star status. Even Teddy Roosevelt wrote that jaguars ". . . will sometimes become man-eaters." Sasha Siemel, the great spear-hunting *tigrero* about whom I did a chapter in *DEATH IN THE SILENT PLACES* (St. Martin's Press, New York, 1981) wrote of a cowboy stalked, killed and eaten by a jaguar which Siemel himself hunted down and shot on the Térére River. My old friend, Eduardo Barros Prado, a famous Amazonian hand and author of *I LIVED WITH THE JIVAROS* and *THE LURE OF THE AMAZON* (printing details unavailable) should have known his apples about man-eating jaguars and did write of them. I visited Prado at his home in Buenos Aires in 1965 or 1966 several times with our mutual friend Eric Gornick, the famed artist and fishing guide of the Argentine Andes. Prado had, as his book claimed, indeed lived with the Jivaro Indians for some time, this tribe being noted for their wars of eternal revenge in which they have developed an involved headhunting ritual whose end product is the *tsantsa* or shrunken head. Eduardo firmly believed that *el tigre* ate Indians and anybody else handy as standard fare. He told me of having shot one *tigre* he was certain had been a man-eater and several more who were highly suspect. Unfortunately, and I have no wish to indicate any incredulity toward a very experienced and fine man, Eduardo did not impress me as more interested in science than sensationalism. He was, after all, a professional adventurer.

If you want to take the time with me, we can scour quite a good many references to man-eating jaguars, but mightily scarce are those fleshier than passing reference. I wish I had asked Siemel, now dead, about his Térére man-eater when I met him, but that was years ago and I did not know that I would some day be a professional jaguar hunter myself or that I would be writing this book. As a personal source of research material, I can only pass on the usual hearsay. To sum up, I believe that the jaguar very rarely turns to man-eating on a casual basis but never, unless in the case of still unrecorded insanity, as a steady diet.

As a man-killer, opposed to a true, unprovoked man-eater, the *tigre* has impressive credentials. Jack him around with a badly placed bullet and a hospital will need your original blueprints to put you back together again. Many people, especially *vaqueros,* (literally "cowers" but the same as our cowboys) are injured and killed by jaguars while working the same herds of cattle a territorial jaguar may consider his own property. I have seen Indians in the Xingu region of Brazil bearing scars from jaguars which they in one or another way displeased.

With any animal as potentially big and dangerous as a jaguar, let's face facts: if you want to be *positive* not to have problems, for heaven's sake stay in Brooklyn!

* * *

As is the case with both the jaguar and the killer whale, who ought to be far better at man-eating than they are, their reputations are largely due to the fact that they are so physically equipped to do damage that most people find it unreasonable that they don't. Another odd and highly controversial example of this combination of characteristics is assembled in the American mountain lion, puma, cougar, panther, catamount or any of the other dozen local names that *Felis concolor* has accumulated from his huge New World range. The aspect of the puma—which I'm used to calling him and will continue here—that is most illogical is the lack of any parallel behavior between himself and his closest counterpart, the leopard of Africa and Asia. They are the same size and strength, puma easily killing deer and other prey similar in size to that of the leopard's. They both climb well and frequently hunt as much from ambush as by stalking and charging their prey. What they do not both do, for reasons of what can only be species "personality" or established behavior traits, is attack man equally. Yes, pumas have killed and eaten people, but so rarely as to almost be discounted. Yet, on a purely statistical basis, this makes a puma more dangerous and likely to attack man than a bull killer whale! Sometimes, the empirical aspects of science don't closely resemble what they're cracked up to be!

One of the most unquestioned incidents of North American man-eating by a puma took place in 1924 in the State of Washington, in which a 13-year-old boy was killed and at least partially eaten while on an errand to a neighboring ranch. This is one of the attacks mentioned by Elmer Keith in his *BIG GAME HUNTING* (Little, Brown & Co., Boston, 1948) although I know Elmer better as the Executive Editor of that great magazine, *GUNS & AMMO*, whose pages I have had the occasional honor to share with him. Oddly, it is also one of the man-eating episodes that turns up in a 1952 paper called *THE DISAPPEARING PANTHER* by O. E. Frye, Bill and Les Piper, the latter of which are the proprietors of the Everglades Wonder Gardens only a few miles up the road from my home here in southwest Florida.

There's little doubt that a good percentage of puma attacks on man, particularly when the victim is not eaten, is the result of one of these cats having contracted rabies. That would be one party I'd prefer not to attend! Unfortunately, the puma is exposed to this disease through its prey and a certain percentage of the population is bound to contract it.

One of the more interesting damnfool things people have done in one or another of the wildlife exhibits or drive-through zoos now so popular, took place a few years ago when a family in a motor home permitted a puma to enter the vehicle through an open (!) window. Whatever its motives, it pounced upon an eighteen-month-old baby and badly injured it before the family's grandmother armed herself with a butcher knife and stabbed the puma to death! I wonder if they had to pay for it?

Although there are several races of puma or "cougar," as it's possibly known better in the American West, where it remains fairly well established, it has been badly mauled as an extended species through loss of range, prosecution for stock loss and the usual problems plaguing large predators competing with man for territory. Some scientists believe that it's coming back to some degree in the east and at least some are still hanging on by their claw tips, possibly as close as one mile from where I write this, at the edge of the Everglades complex. Playing golf two years ago at my club, I found three pug-marks of what I am positive was a youngish puma at the edge of a sand trap. Either that, or somebody around here has a Siamese or Chesapeake I sure don't want to bump into! The print was definitely not that of a bobcat.

There are quite good numbers of both color phases (as I recalled from my hunting days, one has a black nose, the other a pink one; technically there are more than 30 different subspecies due to coloration of fur, noses and what-have-you) in and throughout South America. However, should you for any reason wish to make the record book as having been a puma victim, I would suggest you try the North American northwest, particularly British Columbia or Washington State. Between nine and twelve of the known puma attacks on man have taken place there, certainly more than half.

At this point, again flagrantly admitting I have no idea why these last three species so rarely harm—let alone eat—man, there comes to mind one characteristic common to both the jaguar and the puma which may have gone a long way toward creating a savage and sinister reputation for both. It has never happened to me personally so slipped my mind. The two species being territorial, there have been consistent reports by hunters, surveyors and other outdoorsmen of having been "tracked" or "followed" by either jaguar or puma. This may be a simple act of curiosity; in the face of so few attacks it almost surely is. That this "shadowing" by a potentially dangerous and certainly impressive looking cat might be interpreted by man as stalking preparatory to attacking would be understandable. Pumas and jaguars ought to knock that sort of thing off as it's sure not doing much for their public relations image.

So far as concerns the rest of the felines outside the *Panthera* clan, the lion, leopard, tiger and jaguar; those species with a special bone in their throats permitting them to roar rather than "scream" as does the non-*Panthera* puma, there is no evidence at all that I can find for man-eating activity. Most of the genera is too small, anyway, but even the rare and mysterious Clouded Leopard (*Profelis nebulosa*) and the magical Snow Leopard of Peter Mathiessen's award winning quest to the Himalayas both have clean slates as wanting nothing to do with you, me or anybody else. That probably indicates very good taste. The cheetah, *Acinonyz jubatus* is neither a man-eater nor even a cat, for that matter.

*   *   *

The bright, little plug inscribed an eyebrow of an arch as it pulled soft coils of eight-pound test monofilament off the spinning reel spool and through the decreasing diametered guides of the fiberglass rod. From above, a lock-winged pelican eyed its flight through the soft Caribbean morning air and a tiny black-capped tern butterflyed to the slick surface, picking off a tiny, silver sliver of life whose luck with survival resembled mine with the ponies. With a flat splat, the double gang-hooked plastic lure threw a modest collection of concentric rings, and the throb of its action pulsed through the rod tip as I began reeling in, retrieving the aquatic made-in-Ohio Jezebel in its glittering, flashing dance of ersatz agony through the almost invisible water, spangled but with the chance glint of a turning baitfish.

My mind was somewhere far, far away, crouched in a goose blind or wondering what had really happened to that sable antelope I had tracked for two days the year before; far, far away, permitting it to do what fishing is all about: think about one thing at a time. The lure was a third of the way back to the skiff when something large, strong and unseen sneaked up and ruined my morning.

The strike was purely felonious, a sharp riposte that blended without a blink into a solid *"skreeeek"* of unhappy gears and protesting drab mechanism. And, in that moment, a little slab of my life passed under a new cookie-cutter and would never be the same again. Four feet of Samurai steel sleekness slashed out of the smug surface at a direct right angle to the slant of my line, chest high and straight as a thrown lance. He looked like somebody had hung him there despite his speed, the plug sticking out of the side of a set of jaws I wouldn't have touched long distance collect. It was a barracuda. But not like any barracuda I had ever bumped into. Barracudas, in my experience, made a living ruining marlin and other carefully sewn baits, scaring the hell out of tourists from Vermont and acting like sacks full of coral rocks when hooked while trolling deep water off Bimini or Cat Cay. Obviously, I thought as he screeched off another 150-yard run ending in a towering, jackknife jump, there has been some mistake here.

He wasn't any bigger than what one politely calls a "very nice 'cuda" if somebody else catches him. But, he sure changed my mind about his potential as a game fish, for ten minutes putting on a show that included runs better than a bonefish and an aerial display you don't hardly see but on the Fourth of July in a town with a big fire department and a lot of JayCees. A tarpon would have been pink with embarrassment trying to keep up.

I didn't even get him, either, that mouthful of switchblades eventually touching the heavy leader ahead of the plug and slicing through the tough stuff like an axe dropped on a plate of Jell-O. Since that day, though, I have learned that the barracuda is the most underrated sport fish that swims, especially in shallow water on reasonably light tackle. The late Dean Witter, Jr. and I got into a bunch one day with that great guide, Cal Cochran, out of Marathon in the Florida Keys, and had two hours of what I have to consider the finest fishing I have ever even seen movies of. Barracuda are light for their size, so I guess these three-footers were no more than ten pounds, or so, but on six-pound spinning rigs, it was spectacular. Three times fish jumped not only over the boat, but over our heads! I still fish both giant and baby tarpon on the fly with Cal, throw odd hunks of feathers at nearsighted bonefish and even have sacrificed unblemished white lambs to the gods of the permit (another great Florida flats fish), yet without luck on the fly. But, you'd as soon catch me out on those jade and lapis stretches with an empty cooler as without a spinning rig led off by the relatively new "tube" lure, a section of dyed plastic surgical tubing about ten inches long which is cast either near a spotted 'cuda or near where one ought to be and reeled in as fast as a man can turn the crank! Skittering across the surface like a goosed houndfish, a favorite prey for Jaws, Jr., the strikes are reminiscent of a speeded-up version of the Pearl Harbor attack. Great God, but what action! I'm not going to quit Atlantic salmon or tiger-fish or dorado, make no mistake, but old Mr. B. with the half-smile stays number one in my book. Also, I find it interesting to fish for an animal that can and will, physically and psychologically be delighted to take off your leg with one bite.

Yes, the barracuda *is* a man-eater in the true sense, the only reservation being that he may not eat an entire person at a go, but if you don't mind being processed piecemeal, it sure doesn't bother him! When I first started fishing for barracuda seriously, which is pushing 20 years ago, I had very sincere doubts of 'cuda attack. It is only now that I have spent so many hours in actual observation of these astonishingly fast, deadly-armed and impulse-oriented creatures that I have an idea of what has happened in so many official or semiofficial reports.

There was a period, particularly after World War II, when Jacques Yves Cousteau perfected the SCUBA unit, that the barracuda lost a great deal of esteem as a potential author of mayhem. More and more swimmers were diving and finding out that they weren't being eaten despite close association by curious 'cuda. More and more, the idea was promulgated that what had previously been believed to have been barracuda bite because of shallow water was in fact attributable to sharks; some cases of this undoubtedly being true. But, as the years and the records accumulate, the reality of man-eating by 'cuda becomes undeniable. Perhaps these are, as some delicately put it, mere "accidents." The point is moot if it happens to be your arm, leg, or protuberance more personal, selected by a misinformed barracuda.

Because of the obvious problem men have breathing the stuff or staying submerged for very long periods of time in water, the great predator fish are not well understood. Conclusions drawn by one expert are negated by the next. A Dr. Donald P. de Sylva of the Marine Laboratory of the University of Miami is quoted by Roger Caras in his book, *DANGEROUS TO MAN* (Revised, 1975, Holt, Rinehart and Winston, New York) as, during the writing of his doctoral dissertation of having located "about 33" authenticated 'cuda attacks by 1958. Later, several new ones were recorded. Jacques Cousteau, on the other hand, opines that he has never heard of a barracuda attack. I favor Dr. de Sylva greatly. He was actually researching the matter, Cousteau was not. Also, because of what at least I suspect as attack patterns, unprovoked attacks on a submerged man would be quite rare as even the largest 'cuda would not likely bother to take on something as large as a man when he could not obviously swallow the whole thing.

The Great Barracuda (Sphyraena barracuda) may reach 10 feet and more than 100 pounds—possibly quite a bit more. The teeth of this largest and more common species are so sharp as to defy description, literally as sharp as straight razors. The speed of the animal is so great I would hesitate to guess, except to say that I have seen them move so quickly, when hooked and fighting a mechanical clutch drag the equivalent of four or five pounds or more, that they were too fast to follow over a clear sand bottom with the human eye. I've got pretty fair eyes, too. I would not quote a miles-per-hour guess, but I once found the fresh half of a bonefish on a flat in Abaco that could not apparently outrun a 'cuda. There are no barnacles growing on bonefish, either, but I would say the 'cuda is much faster, although there's certainly room for other opinions in this matter.

A bite by a big 'cuda, say a six or eight-footer would produce the type of wounds once described by the *Journal of the American Medical Association* as "characteristic of attack by large specimens." The wounds indicated are "total amputations." I would clearly agree with Dr. de Sylva that the odds of barracuda attack are fractional of those of being taken by a shark. Still, the reason the 'cuda is included here is that it does happen.

The general feeling about the circumstances of man-eating, or at least the partial man-eating of the Great Barracuda is that the fish does not realize that he's attacking a portion of a much larger object when he takes a hand, arm or leg. In my experience, barracuda see very well, provided that water is clear, one reason many attacks are in roiled or dirty water, close to the wave line. On the clear flats, a big 'cuda can be tougher to hook than a brown trout in a British chalk stream in August if he has a chance to look the lure over quickly. By all my observation, a 'cuda is generally either hunting or he isn't. If so, a lure that moves fast, forcing him to make up his mind to attack before it "escapes" is the most likely to bring a strike. I think of this as "impulse" attacking; the fish was just ready and, one look at his streamlining shows clearly he wasn't designed to hang around counting his options. He may see the lure as a houndfish or a minnow the same way he perceives a foot with an anklet or red toenail polish: food. Same goes for a hand with a ring, an arm with a bracelet or a neck with a locket or other shiny disk. Sharp, erratic movement excites barracuda. Sharks, incidentally, also respond to this.

My only comment to the many skin divers who have laughed at the idea of 'cuda attack is that, submerged, taking basic precautions, they're probably very unlikely to be bothered. But, that's up to them. When an eight-foot 'cuda comes nosing by and shows an interest in yours truly, however much is curiosity, I'm swimming very slowly and deliberately to the surface, getting quietly into my boat and going home.

I remember very well, in my young and foolish days wading for bonefish somewhere or other and turning around to note three barracuda, who looked like something Weyerhauser would cut down, in the knee-deep water about two yards behind me. As I waded along, they followed, showing, beside a need for orthodontia, an almost hypnotic interest in my white tennis shoes as they puffed along the sandy grass flats. That they made me nervous would be a mild description of my sentiments, which were identical to being closely followed by a pack of 300 Siberian wolves who had spent the last couple of months on the Scarsdale Diet. Yes, "nervous" would cover it quite nicely.

My luck that day was excellent, as I saw not a solitary bonefish and have since considered the question of whether the fish juice and the thrashing of fighting one might have triggered one of those big bastards into trying a slice of 10½-D Keds and the contents thereof. The lodge manager assured me that such "following" incidents were fairly common and there had never been an accident. I also happened to ask him if he ever waded the bonefish flats. He definitely did not. "Too many bloody 'cudas, sir," he told me confidentially.

Ah, well, Allah protects the ignorant.

Leon Parson

CHAPTER NINE

CANNIBALS

It seems rather novel to exist in a civilized state—or, at least one advertised as such—and contemplate that we are surrounded closely, in fact, overrun, with the one animal that, over the untold generations, has been the most deadly enemy of man, the greatest killer and eater of our kind that has ever existed. As would be consistent with most of the ironies inherent with our species, of course, that menace is the most unlikely of all. When political cartoonist Walt Kelly, creator of the *"POGO"* strip had one of his characters declare "We have met the enemy and he is us," he really hit the nail on the head; in fact, hitting each other over the head for fun, profit and culinary gain as far back as we have been able to trace, seems to be one of our most practiced and deft talents. There are those today who will still maintain that man, the little hairless, puny-toothed, slow frog-muscled brute who has placed himself so rudely in charge of just about everything worth controlling, is not really a killer, the ultra-evolution of the homicidal biped, but I think they're whistling in the dark. It's not a pleasant concept to contemplate, and we know that historically we contemplate unpleasant matters badly; look for example, at the public outcry against that evolutionist crackpot, Darwin. Yet, more men have died, directly or indirectly, by the hand of their fellow men than any other cause. And, usually for the noblest of reasons. . . .

Man as an eater of man is one of the really spooky genuine subjects left to those of us who mix metaphors for that most lofty of capitalistic ideals—gain. If you haven't noticed, there is a weird cult that rather stands out on its own on the plain of humor, some looming manadnock of off-beat amusement that keeps people like Charles Addams and Gahan Wilson with their baroque, black (and excellent) cartoon humor, solvent. If, beside good old sweaty sex and the constant stream of epics of the American West, there is a subject more exploited than vampires or werewolves, it comes not instantly to mind. We seem to have some off-beat propensity, as a species, for liking our collective spines tingled. There's nothing really even remotely funny about a missionary being boiled to death, although there must be dozens of jokes on the subject. Cannibals are traditional stars of golf course humor, usually with prep school level sexual overtones. But, why? I suspect that even today, when most of our problems are wrapped around matters a bit more realistic than somebody rearranging our general profile with the thigh bone of an antelope and dragging our carcasses home to the equivalent of the microwave, the tiny, niggling ancestral thought of Man the Prey still lurks back there somewhere under our—speak for yourself, and I do—receding hairlines. Actually, if you wish to take the concept right down to grass roots, the only reason you're here and able to read this book, the only reason you exist, is that you come from an unbroken line of forebears, stretching unimaginably far back into the ground fog of creation, every single one of which, without fail, managed to procreate before dying, whether the cause was disease, tigers, war or ritual sacrifice. So, rejoice! The odds on your being here in the first place are so lousy as to be almost laughable!

I suspect that very few of us realize how close cannibalism and the rites attached to one or another form of the practice still are to us today, smack in the middle of the age of Skylab and Superbowl and the self-declared moral majority. But, rather than jogging roughshod in overgenerality over what is a delicate subject, thought by many to be a quaint relic of the steaming Tasmanian or African jungles and rocky, lush tropic isles of the Seventeenth Century, let's take a quick look at the practice and theory of cannibalism which in itself, gives a hazy mirror-image of a part of us we really don't want to see.

I am not nearly bright enough to have the slightest anxiety of this chapter sounding like a textbook, so let us sitting well in order, smite the most original basis of one person eating another; a practice, incidentally, which undoubtedly spawned the term which refers to one being "in good taste," an anachronistic paleophrase still found in occupied Connecticut. Man started out on the same basis as anything else composed of meat. He tastes good. In some cases he is reputedly not very easy to clean, but then, he's not too bony, either.

The earliest evidence we have that being a practicing cannibal didn't necessarily make you a bad person is pretty well synchronized with the strata and other dating methods that take us back to the first of our kind so fortunate to be called "man" or at least to be considered in the running. They've got a slew of species and subtypes, ranging from the earliest types in East Africa and South Africa, mostly Pithecanthropus, Sinanthropus, and what-have-you (you have plenty!) to the earliest—other than the australopithecines—form of "man" we recognize, Pekin or Peking Man. Four hundred thousand years ago, he spent his spare time eating the brains of his pals or enemies, perhaps both. Because of the deliberate marks on the bases of skulls, widened at the juncture of the spinal chord to facilitate removal of the brains we know he was a cannibal; unless, of course, he used the brains for shoe polish or something else more palatable to the sensitivities of people such as Joan Marble Cook who, presumably outraged that our heritage might be felonious, produced a truly uplifting work called *IN DEFENSE OF HOMO SAPIENS* (Dell Publishing Co., New York, 1975). It is a rather optimistic refutation of the work of paleoanthropologists and thinkers such as Robert Ardrey, Desmond Morris, Lionel Tiger, Robin Fox and others who would tend to keep their backs to the wall were they to encounter one of our more remote relatives. (Don't take it too hard, Joan, I spelled your name right!)

Since the only thing we really know about the cannibalistic practices of the very early hominids was that they qualified for the title, thus, let's just keep it simple and—right or wrong—lump this type of behavior (which is *not* necessarily anti-social) under the most obvious form of cannibalism, that for food value only.

Things start to get a touch involved about 75,000 years ago. This is hardly recent, but a drop in the proverbial if you're talking about the possibility of some protohuman remains being in the area of five millions of years old. I suspect that most people reading this book are of an age to recognize their high school exposure the so-dubbed Neanderthal Man, from its discovery in Europe's Neander Valley, and which has a silent "h." So, it's *"Neandertall."* Whatever. These were rather squat, presumably ill-tailored folk, with halitosis, who are the darlings of the modern funeral director, because they apparently came up with the idea of burial in the first place. Cannibalism being the order of the day, although with most probable mystical overtones, it's my personal guess that burial was essentially a product of hiding the body of a clan member so it wasn't found and eaten by the bad guys. The Neanderthal chaps were what the books will tell you were likely exophagists who ate only friends or/and relatives. Confusing, what? Well, here's where it really starts to get involved. . . .

As far as we know, the first *surviving* evidence of ritual in relation to body disposal, whether through cannibalism or otherwise, has been recorded in this early Neanderthal epoch. The whole idea, albeit on a very wide scale, centers around the central hub of the concept of *soul*. This, of course, has caused us more problems through the ages than any other philosophical/theological stew we have ever dreamed up, because as soon as we had one or another idea of *soul,* the logical procession of starting to create gods in our image was a natural progression. Put the whole bloody thing down to too much spare time, I suppose. When food, even human meat, was abundant, we started to develop specialties, such as one man in a clan or family group doing nothing but make projectile points while his brother specialized in setting snares. While somebody was squatting here or there, they got the spare time to wonder what in the world was going on when they slept; especially when they dreamed. This was active soul "proof" since one dreamed of being else-where than on one's sleeping mat and of doing activities, many of them undoubtedly quite lacivious. The concept of dreaming while in sleep led, probably within the first ten minutes of contemplation, to the idea that, if one dreams while in the unknown function of sleep, certainly one must do the same when dead. Obviously, death was (and in literature still commonly is) another, probably advanced stage of whatever sleep is. Therefore, *soul* is still wandering here and there, although now it's called *ghost,* or whatever the equivalent might have been in the local jargon.

The sequential whistle stop of this whole business comes down to the basis of what is today religion, basically a relationship of an unknown power (gods) to an unknown and misunderstood portion on one's self; the "dream body," soul or ghost. Since nobody quite knew what went on or happened to the "dream body" after death, let alone where it lived in the body, the body itself became—despite imperfect logic—important as an edible talisman as much and more than a food source, be-cause one could capture anothers force or soul. Eat your enemy and you absorb his bravery. Eat your father or mother, with appropriate ceremony, of course, and ab-sorb their knowledge and strength. Keep them with you. Be *one.* If you've got some question about this, go to the library and wade through the biggest collection of the dullest goddam books you ever did see, but I suspect that the concept is pretty sim-ple, even as expressed here.

Okay, so far we've touched on "natural" cannibalism, in which one is added to the larder because there's no reasonable philosophical or religious reason why he shouldn't be, as well as "magical" cannibalism, which is a part of the next classifica-tion if not indeed a proper subset: "religious" cannibalism.

Most Christians who partake of Holy Communion do not have a tendency to consider themselves as participating in a cannibal rite, actually as a Catholic and symbolically as a Protestant. What is forgotten is that Christianity as we know it today, as well as many other religions, is not exactly the same sociophilosophical entity as probably perceived by the founders, such, in the matter of Christianity, as Christ. Even in my lifetime there have been changes in Catholicism that would have gotten otherwise pious folks fried a few centuries ago. I don't happen to be a Catholic, but religion has always interested me generally to the degree that I'm aware that the American Mass is now in English, that there is no confession, that meat is now permissible on Friday and that there are probably other changes that would make the Disciples of Jesus wonder if they had the right address when attending a modern service! Whoa! No rights or wrongs: no judgments; simply observations on what I perceive to be the historical facts. No offense intended, I hope none taken. It's just that I'm reminded of a very beautiful African evening about a dozen years ago, in Zambia, around a mopane wood campfire no brighter than the stars over my old gun bearers, Silent, Invisible and myself who were scouting elephant and camping on the track of what looked to be a whopper.

Silent has always been one of my favorite people. Older and with an extraordinary sense of natural dignity despite his cast-off bush rags, we spoke often of the philosophies of our two worlds; a fair feat considering that he could not conceive of what an ocean might be or that one could not go to "Amelica" by train from Lusaka. Heaven knows what he perceived my short comings to be! On this particular night, with his brother, Invisible, also smoking quietly at the fire, he asked me to explain the Christian religion. I cannot conceive of any clergy not having me up for heresy as I tried to put Immaculate Conception into terms a couple of most sexually explicit bush Africans would understand, but I did my best.

Silent and Invisible were Awizas, although we were in Senga country, but both locations were traditionally, to some degree, of cannibal heritage. In fact, a common tradition still seriously held that one of the local Paramount Chiefs owed (through the obligations of descent) another one a prime human hand, which had apparently been borrowed back when such delicatessen services were permissible. The palm of the hand, incidentally, was once considered the best of cuts in these parts, not far south of Congo. While I'm at it, I might as well mention that several denominations of Protestants offered free tee-shirts to the Africans if they would be baptised and join their particular sects; one skinner I hired had four, all dutifully earned by instruction in one or another brand of Jesusism.

As the night went by (it probably wasn't more than about nine) Silent went, well, silent, considering my Immaculate Conception explanation with the natural courtesy not to guffaw out loud. (You think I'm kidding; you try it sometime!) At last, and I knew there was something quite heavy on his mind, he softly clapped his hands for formal audience indicating he had something important to discuss. It went this way:

"*Nyalubwe*," he started in his soft, reedy voice, using my highly personal African name, meaning "Leopard," "tell me a thing."

"*Eeeeehh, medalla*," said I in dialect, agreeing.

"Tell me this, *Nyalubwe*," he said, dragging at a cigarette he had just rolled:

"Your fathers came to this place, the *WaIngisi*, the English, and told our fathers they might no more eat of man. This being so, why is it that you eat your god?"

At first, I couldn't figure out what he was getting at until, like a club behind the ear, I realized that he was asking about communion. My explanation wasn't very hot, either and I dare say that there might be a bit more work done on the part of well-meaning missionaries in the matter of explaining the more difficult portions of religion to the heathen than there has been, and I don't just mean African tribesmen. I have yet to find the cleric who can explain to me precisely what the Holy Ghost is, but, then, I'm clearly not very bright.

What I'm getting at is that, in the case of Catholicism, since the year 1215 and the Fourth Lateran Council, when the priest conducting Holy Communion intones the Latin *"hoc est corpus meum"* or "this is my body," the body and blood of Jesus Christ are no longer symbolically represented by the bread and wine but, under the doctrine of transsubstantiation, theologically are the blood and body of the Saviour. If you don't believe this, you, friend, are guilty of heresy. Christ made this clear through at least Matthew, Mark and St. Paul who, despite the fact that eyewitnesses are statistically the least reliable, do seem to agree upon the statement of Jesus that the bread was his flesh and the wine his blood. Whether he literally meant this is still the subject of great conjecture, but it certainly points up the tremendously antique origins of what is even today an ancient religion at a time when it was simply a thread of belief. Of course, the concept behind the words of Jesus, if indeed, a mortal may guess at them, must have been in the ancient tradition of endophagy, the sharing of the most sacred part of a person, his blood, his flesh, his self-essence, as an act of brotherly love. It remains as a most fascinating anachronism, particularly because in the concept of Judaeo/Christianity, the eating of flesh and the conflict in both philosophies is polar.

If there is an area of confusion greater than the official Catholic Church status on anthropology, I haven't found it despite a great deal of looking into the problem. Here's the way things stack up: Christianity, for centuries looked upon cannibalism as a crime worse than murder (as did the Jews) for one basic reason. It was, of course, the expectation of Judgment Day when, Hallelujah, the earth would give up the corrupt and so would the sea. Important Jewish and Christian clergymen were buried in the clothes they wanted to appear in when the trumpet sounded and, obviously it followed to the medieval mind, no body: no Judgment Day.

Yet, we find, as did the survivors of the horrific crash of the Uruguayan airliner that came to an abrupt halt against an Ande, the survivors of which stayed there, keeping alive for more than 70 days by eating each other's bodies in 1972, there was, among those who made it, a strong religious feeling of having done "the right thing" by cannibalism in terms of "God." A precedent of their act was stated by spokesmen to be the Holy Communion. The clergy, noting that there was no actual rule against cannibalism in Christianity while *in extremis* confirmed that they did not have to "forgive" the cannibals but merely assure them that they had done nothing wrong. Personally, although I would not like to make the actual decision, I see nothing wrong by staying alive at the expense of the already dead. Starvation must be a singularly unpleasant way to die. Besides; who would *dare* judge one when one was forced to make that decision?

Well, there's religious cannibalism, at least one symbolic form of it. The Mexican Aztecs were pretty handy in this department for, if nonparallel reasons, at least similar resulting deaths. As is well-known, prisoners of war were the primary prey in their system of keeping the sun going, but the cannibalism itself was highly formalized and really a subfunction of the entire religious service. Of course, some of Cortez's men who became most direct participants, in view of their fellows, might not share my flippancy. Their hearts were torn out, still beating, and their bodies kicked down the pyramids. In the case of the Spaniards, however, all were eaten with the customary local condiments, tomato and peppers. Not all Aztec sacrifices were eaten, but imported filet of Spaniard was too much to pass up.

In passing, it's worth noting that the feeling of paleotheologians seems to be that the modern customs of dropping small donations in the poor box and the purchase and lighting of candles before the altar in Catholicism is the vestigial, pre-Christian remnant of the sacrifice of burnt offerings, human or otherwise.

Perhaps I have oversimplified in the mutual inclusion of religious and magical cannibalism, but, if viewed from any distance at all, religion is itself a form of socially acceptable magic; or at least surely once was. The whole ball game properly falls under the heading of "sympathetic magic" anyway and, since this is not a sociology textbook, my better instincts advise that enough has been said already on this facet of the matter.

The elements of cannibalism that fascinate me more than the technical rules and definitions of the activity itself are those adventures associated with cannibals that survive through reliable literature. Cannibalism, or at least the eating of man (or any other species eating members of its own) under that title, is, curiously, one of the first words to derive directly from the New World. When the Spanish, as well as Columbus himself, began poking about the new islands they quickly found the local terrors of the area to be the Carib Indians, a charming collection of people-eaters who, painted completely in crimson, had almost conquered the Lesser Antilles and were fast making inroads into Jamaica and Puerto Rico. Columbus, upon sailing past the island of Montserrat, was advised by other natives that the place was deserted, every inhabitant eaten by the Caribs. More than 200 years later, in 1694, a Father Labat wrote of having been presented with the arm of an Englishman (Ugh!

(167)

An Englishman!) killed, smoked and preserved by a party of Caribs newly arrived at Martinique. The Englishman was one of six people recently killed on the isle of Barbuda, about 30 miles from Antigua. The fact that the arm was slow-smoked prompts me to also mention another purely New World word. This was originally derived from a Carib term for curing or cooking meat—usually human—over a lattice built over a fire. French pirates, who found the process handy, adopted the Carib word, *boucan* and themselves became *boucaniers;* thus, buccaneers. The Spanish called the latticework that held the meat over the fire *barbacoa* and this ended up with the ever-popular *barbecue* in English.

It was the corruption of the tribal name of the red-painted, man-eating Indians themselves that produced the term *cannibal.* "Carib" was twisted by the Spanish first to "Calib" and then to "Canib." From there, the term "cannibal" was home free in our language. It must have been a busy little area, the "Caribbean," about 1500, to leave such bloody spoor of linguistic tracks that linger even today!

Africa was always traditionally a great spot for cannibalism, and in fact if the rest of the Nineteenth Century world had known how widespread the practice was, even the missionary societies might have been surprised. As a practice, cannibalism was well established long into even this century and, without question, can still be found in more remote areas, if not on a natural, then on a magical basis. During the Kenya Mau-Mau "Emergency," many of the "oathing" ceremonies, so hairy as to preclude insertion even into this book, specialized on the subject of man-eating, but to say that the degeneration of what was at least a colorful, human fringe activity, honest cannibalism, was certainly taken over by some minds with a lot of spare time on their hands! The unspeakable cannibalistic distortions of many of the oathings later recorded by pals of mine who fought the primarily Kikuyu movement, as far as I am concerned, literally defy reproduction here. Still, three men whom I have studied a great deal all had extensive experience with cannibals; in fact, if you include the third man seriously, he *became* a cannibal beyond any other reckoning of the matter!

In my second book, *DEATH IN THE SILENT PLACES* (St. Martin's Press, New York, 1981) one chapter was devoted to the exploits of P. J. Pretorius, a South African adventurer, ivory hunter, naturalist and soldier of fortune who almost single-handedly was responsible for the German defeat in the East African Campaign in World War I. Before the war, he got into one hell of a mix-up with Congolese cannibals who were killing and eating his men, finally attacking his camp in an all-out assault which Pretorius defended with a shotgun with such a volume of fire that the solder holding the barrels of the gun together got so hot that it melted, leaving him and his few survivors unarmed but for the white man's pistol and a few cartridges. I won't ruin the tale of his attempted escape from the butchering knife for you, should you wish to read it as I wrote it earlier.

A tale I didn't write concerns one of my favorite African characters, George G. Rushby, the same chap found earlier in this book in connection with the Njombe man-eating lions and therefore, who needs no reintroduction. George was hunting the Mbayaki area of the Congo in Pygmy territory for ivory and had been warned by the chief of a Pygmy village (Pygmies were, in so far as I have been able to research, never associated with cannibalism; interesting because they *do* eat monkeys and gorillas) that there were Bantu (negro) cannibals in no short supply locally. After a three day hunt, Rushby managed to kill two decent bull tuskers in a swamp, camping on a high spot some 300 yards from the carcasses to dry out and remove his accumulated leeches. To his surprise, despite the fact there were no vultures to pinpoint the bodies, about 30 Pygmies had found the bulls and were cutting them up.

This is an interesting "phenomenon" frequently reported by hunters new to Africa; that being the sudden appearance of local tribesmen almost instantaneously after the killing of a lot of meat such as an elephant. This is not a function of telepathy or some jungle telegraph (My God, Quimby, the drums have stopped!) rather, the simple fact that a party obviously hunting elephant are shadowed at a reasonable distance by locals who appear after the kill and normally most politely ask for meat, which is generally granted. Does the safari client think the villagers spend all their time wandering around with a collection of meat baskets and more knives than a cutlery shop? In Zambia, I made a deal with the nearby villages that, if they would not "shadow" us and possibly spook our game, I would faithfully send a Land Rover to advise them of any kill of a big animal with more meat than our camp would use. It worked: nobody appeared like magic again. I suppose it was too much trouble to follow us when they knew they would be notified. Alas, another legend is flushed!

Some hours after finding the Pygmies drying the meat, about a dozen very bad-assed looking blacks arrived at the carcasses and asked if they could have some *nyama*, too. Rushby, noting that they were armed with muzzle-loaders, poisoned arrows, spears and such wasn't hot on having them around, but consented. However, he cautioned, although there was meat for all, any fighting would be an excellent reason for him to shoot them all. There was no squabbling, and later the headman of the blacks wandered over to George's camp and chatted with him. According to the man, his small tribe didn't see eye-to-eye with the French, and their favorite intramural sport was sniping the mercenary Senegalese troops which roamed the jungle edge. The black added that George need not worry as Rushby was well known in the country as being an English hunter and not a Frenchman.

Rushby twisted a smile from beneath his barroom smashed nose and asked whether the local cannibals made such pleasant distinctions. With a matching smile, the tribesman flatly admitted that he was himself a cannibal and what of it?

(169)

Philosophizing together that quite probably nothing that is killed for meat is especially fond of the process, Rushby asked the man out of curiosity whether he preferred game or people as a staple diet. Describing the fact that human flesh is nicely marbled, rather the same property found in prime beef, varigated with streaks of fat through the lean, the cannibal advised George that human meat behaves much more actively in the stew pot than does game flesh. Being much "lighter" than animal meat, it bounces and bumps around as if it were alive. Charming! Oh, yes, decided the cannibal sagely, human flesh was by far the best.

At this point in the conversation, Rushby made a mistake that was to cost at least one person his life. After asking as to the texture, and being told that it varied from individual to individual as well as presumably from cut to cut (which, to me, gives a very truthful ring to the conversation, were one to have any doubt of its authenticity) Rushby rather impetuously asked for some human flesh insofar as he had given plenty of elephant meat to the at least part-time cannibal.

The man agreed at once. Telling George that he happened to have quite a lot on hand at the moment, on the drying racks back at his camp, he would be delighted to send over a nice piece as soon as he got back. Rushby, who wanted to see the cannibal camp said that he would come by the next time he was hunting in the area and pick a piece out for himself. Swell, reckoned the people-eater. "You will be very welcome. I will select something special for you."

The very next day, the two elephant carcasses having been stripped by both the Pygmies and the cannibals, the little people led Rushby and his porters to another hunting ground, passing through the cannibal camp on their way. Here, George saw, were numerous drying racks over slow fires, heavily laden with meat. It didn't take a second look to make certain that the meat was not that of either gorillas or monkeys. Fascinated, George picked a juicy chunk, said thank you very much to the head cannibal, and went on his merry way to set up his own camp farther on. After examining the meat he gave it a quiet burial.

There being quite a bit of elephant sign locally, Rushby decided to start hunting intensively the next day. Stopping by the bearers' section of the encampment, he immediately noticed that one of the porters had a woman along. In any serious professional hunting camp, this just isn't done and all porters know it. Women are pure trouble under bush conditions, I can personally assure you, having had one of my men get drunk and make a pass at another's wife, who, unbeknownst to me, was visiting. When the offender passed out, stone-cold helpless, the husband calmly, carefully beat his skull into a crushed pouchful of mush with a length of firewood. As I recall, this being 1975 in Rhodesia, he only pulled a few months for the offense. If you want to steal somebody's gal, don't try it in Africa, or the *Macomber Affair* may look like a "Sesame Street" rehearsal.

Rushby immediately fired the carrier responsible, but the man was so contrite, promising to escort the woman (his wife) back to his village and immediately return that George relented but promised to dock the man's pay for the two weeks he would be gone on the round-trip walk.

With the numbers of well-tusked elephants in the area, the couple had hardly left than Rushby and his men put their minds back on hunting. After three days, however, George was surprised to see the husband returned to his duties. Asking the headman of the carriers how the husband managed to get to his distant village and back so quickly, the man seemed confused.

Feeling his heart sink, George listened to the headman explain that he thought the *bwana* understood. The woman had of course been taken nowhere but a short distance out of camp where her husband had murdered and butchered her. Rushby hadn't realized it, but he himself and the other men had been eating her body for the past two days. With his gorge rising, George understood that, when he had taken the first piece of flesh from the cannibal, all his men thought that he was also a man-eater; just one of the boys. They would have thought him mad to know that he had buried the meat rather than having eaten it as they presumed.

Depressed to the depths of his soul at having caused the woman's death as well as having eaten her, even not realizing it, Rushby packed up the safari and headed out of the forest, finished with the horrors of Congo.

That George Rushby ate human flesh for two days and apparently did not even notice it seems to indicate that it must be at least passable and possibly as good as advertised. Just how good it is depends upon one's source of opinion, and of those available, there is one, a rather occult-oriented gentleman who seems exceptionally well-informed, if ever there was one. First put onto his book about the spookier side of Haiti by my late brother, Tom, back in the 1950's, I read *THE MAGIC IS-LAND* by William B. Seabrook (George G. Harrap & Co. Ltd.; late 1920's London) and also his *ADVENTURES IN ARABIA,* which I think was Seabrook's first effort and not nearly so cultish. But, by the time Seabrook came up with *JUNGLE WAYS* in 1931, he was what the kids of today would definitely call "far out." You see, *JUNGLE WAYS* is basically Seabrook's story of his trip to West Africa to try cannibalism for himself. Among the Guere of the Ivory Coast he was to achieve his wish. Boy, was he!

After three subchapters of this rather odd tome, a mixture between semiscientific observation of tribal customs and the standard, practically expected humor of the thirties in which, almost to formula, one of the chief cannibals wears a French fire helmet, Seabrook ingratiates himself to the extent that he is presented with two pieces of human flesh for his experimentation, a rump steak and a small loin roast. I've given a lot of thought to whether or not I believe the following account and have decided that, essentially, at least, I do. I've read enough Seabrook to trust that he was as weirdly fetish-fascinated as his writings indicate. Also, I've checked some facts through no little trouble and find that some of his smallest observations on remote tribes are corroborated by studies that took place by accredited scientists after he wrote his works. I suppose that what I'm saying is that William B. Seabrook, Esq., was not only a cannibal, but collected and tried different recipes for human meat. He was a human gourmand who described in detail his first cannibal meal . . .

"It was the flesh of a freshly killed man, who seemed to be about thirty years old—and who had not been murdered," wrote Seabrook. I have pondered his comment about "having not been murdered" and am not sure what he meant. I clearly don't think he meant that the man had died of disease, since he said "killed" which would only leave accidental death or possibly Seabrook's own mental differentiation between death in war or by human sacrifice. Perhaps the man was a suicide. Whatever, the line practically drips with innuendo and I'm not sure what the point is. Along with the "sizeable" rump steak and the roast though, were included the ingredients for rice stew with red peppers.

Seabrook described the raw meat as, "in appearance firm, slightly coarse-textured rather than smooth. In raw texture, both to the eye and to the touch, it resembled good beef." Seabrook reported the color to be reddish, neither pink nor grey like mutton or pork. His observations are somewhat at odds with those of George Rushby's cannibal pal who spoke of the fat marbling: "Through the red lean ran fine whitish fibers, interlacing, seeming to be stringy rather than fatty," said Seabrook, "suggesting that it might be tough. The solid fat was faintly yellow, as the fat of beef and mutton is. This yellow tinge was very faint, but it was not clear white, as pork fat is." In this state, after several good sniffs, Seabrook pronounced the human steak as having a clean odor typical of any large, freshly dead and cleaned animal. Sounds great, hey?

" . . . because I was anxious to get the clearest first impression possible of the natural meat, and feared that excessive condiments would render it inconclusive," Seabrook decided to prepare the portions of the gentleman in the simplest manner, "as nearly as possible as we prepare meat at home." (Hi, Bill? Listen, I'm *really* sorry but we just can't make it tonight . . . ") So, he spitted the roast and started to grill the steak, remarking that it took longer than it would have in London as his wood fire wasn't as hot as the gas flame he was more used to.

Seabrook makes the interesting comment that the cooking produced no peculiar or particular odor, rather as one might recognize by smell the preparation of fish, pork or beef. In fact, at least by Seabrook's lights, both ungodly cuts were beginning to look mighty appetizing. With a knife, he cut into them as the roast browned and the surface of the steak began to char. "The fat was sizzling, becoming tenderer and yellower. Beyond what I have told there was nothing special or unusual. It was nearly done and it looked and smelled good to eat."

Determined to treat the meal in fair comparison to any other, the epicurean Mr. Seabrook determined to serve it in precisely the same way as any other steak or roast. No gumming of tiny morsels fit for goldfish for him! He would take his meat with rice and a bottle of wine, apparently omitting the red pepper customary as a seasoning in the Ivory Coast as it would mask the flavor of the gentleman he was ingesting. The most irritating aspect of this entire portion of the passage was the fact that he did not identify his choice of wine! We must, therefore, make a reasonable determination, based upon the hints of Mr. Seabrook, as the best choice for a rump roast of 30-year-old Ivory Coastian, presumably not overtensed muscularly, but still a bit stringy.

I am light years from being an expert—if, indeed there *are* any beyond their own opinions, in the mellowing art of the wine advisement. I have, however, had some experiences that I do not believe many others have been faced with. What would you recommend, *Monsieur* Capstick, with the *guanaco a la Milanese*? I have been asked this in Argentina by a French client contemplating an *asado* of New World camel. (Obviously, a nice Mendoza red such as *Don Valentín* is the clear choice for *guanaco*.) I've had a variety of the very flexible South African wines with such standard fare as rhino or hippo, dependent upon the particular cut. Christ, man, if you don't know what to order with *capybara*, the world's largest rodent, slow-cooked with the guts left in as a specialty in Belize, then you just don't know your way around. What do I recommend? Plenty of whisky, well before dinner, and an absolutely inactive imagination. *But*, what a fantastic dilemma Seabrook was in, presuming he brought something of a reasonable cellar with him to the Ivory Coast! What *is* proper with rump steak *Guere* or Loin of Negro flambe? Well, I suppose we had best check first what the meat turned out to resemble once prepared . . .

"It was good to eat," advised our local expert, Seabrook, "and despite all the intelligent, academic detachment with which I thought I was approaching the experience, my poor little cowardly and prejudiced, subconscious real self sighed with relief and patted itself on the back. . . . I took a good big swallow of wine, a helping of rice and thoughtfully ate half the steak."

What was it like?

Veal. Not young, tender, baby veal, but not beef, either. In fact, it was so like slightly older veal that Seabrook claims that the average person would not have been able to tell the difference without some hint. In fact, I believe I could safely condense Seabrooks findings under the single adjective "bland." He clearly says that it was not the pork-tasting "long pig" human meat of the Melanesian Islands. Even the roast, of which he ate a center slice, was "tender, and in color, texture, smell, as well as taste, strengthened my certainty that of all the meats we habitually know veal is one meat to which this meat is accurately comparable."

I have never known a self-admitted cannibal save one, and he, like Seabrook, was white. Although he is now dead, his identity shall remain private. I knew his background and experience as a famous professional hunter in an area where the ritualistic aspect of cannibalism was still reasonably common during his tenure and when he told me he twice was obliged, in the interest of interracial good manners to eat human flesh, I believe him. I did once, before I was married, date a lady in New York City who, if such has not already been the case, will almost surely be convicted of eating her young; figuratively, at least.

But, this does not bring us back to the proper wine for a meal of human meat. Personally, I do not believe in the rosé wines any more than I believe in bisexuality. Either you *is* or you *isn't*. For my money, I'd choose a nice Graves, a good year of Chateau Olivier as my wine, not forgetting that if one preferred the loin on the rare side, possibly a bright, young Beaujolais might be interesting.

There is one aspect of cannibalism that I have been unable to satisfactorily resolve or even determine the importance of, such detail being the filing of teeth, and the supposed importance thereof. The most complete recent book on the subject of cannibalism I have been able to find mentions nowhere the idea of sharply filed teeth, although this symbol of the eating of human flesh—if indeed it is so—is almost universal. I cannot give any light on this supposed practice and am starting to wonder if it wasn't a defensive threat measure more than anything else. After all, the ladies with such huge platter-lips from the Bangui region were deformed in this manner to make them so ugly that slavers wouldn't even bother with them. Perhaps people filed their teeth so they would appear more aggressive than prey-like.

America, considering its blushing relative youth, has produced some very enterprising cannibals, not the least of which was the recent hero of the Hollywood opus, *"Jeremiah Johnson."* Tinseltown is, of course, not famed for its adherence to the hard-core truth of history or exactly unyielding portrayals of all personalities, but to make a romantic hero out of what was almost surely a homocidal, reclusive maniac cannibal like Jeremiah Johnson and go so far as to portray him by Robert Redford is even beyond the pale of the land of make believe.

The American West in the first half of the last century was not especially conducive to the recording of precision history; thus we don't know all the details about Johnson except that he was mostly a loner who, like most of his contemporaries, the mountain men trappers, took an Indian woman into a very casual arrangement. Legend has it that she was pregnant with Jeremiah's child when killed by a band of Crow tribesmen. Apparently, Johnson was quite fond of her or perhaps her death had nothing to do with the behavior of Jeremiah after all. Whatever the reason,

Johnson took up roaming Crow country for years, specializing in killing any member of that tribe he could catch in the sights of his .50 caliber Hawken rifle or under the blade of his tomahawk. Dead or alive, Johnson would then slice out the liver of the Indian and proceed to eat it on the spot. One authority claims that his total of Crows reached an incredible 247 people, although whether that was Jeremiah's claim or a genuine count I wouldn't want to say. I rather doubt anybody followed him around keeping score. Still, it seems safe to give Johnson his due: he was one hell of a successful cannibal.

In August of 1846, a very different scenario developed when a wagon train of 84 men, women and children led by one George Donner decided to try a short cut through the Wasatch Mountains of Utah on their way to California. After a terrible struggle through rugged wilderness, it looked in retrospect like not a very good idea. When faced with an 80-mile stretch of searing, salt desert, the move was pretty clearly a downright poor idea. Carrying all the water and grass they could, the party still lost most of their draught animals to thirst or escape.

Somehow, the remnants of the Donner party made it across the 80 miles into Indian territory. Five men were already dead from murder, accident and simple inability to keep up. One man was banished for his part in one of the slayings and another went on ahead to Sutter's Fort for extra food, returning with two "tame" Indians to help out. But, it was already the middle of October and the winter was starting to sweep in early, turning the snaggled Sierra Nevada into its Spanish namesake: The Snow Covered Range. The Donner train was caught just short of the summit and suddenly it was desperation time around the old corral.

Making do with what they could, the party was lucky to find one old, abandoned cabin and built ramshackle huts near what is to this day known as Donner's Lake. Through stupidity and inexperience, the horses and cattle were permitted to wander off, leaving the greenhorn families not only with the possibility of freezing to death but with the looming prospect of starvation. After more than two months had gone by, in desperation a group of seventeen people tried to cross the summit on makeshift snowshoes; ten men, two boys and five women. It was cold. Savagely cold. As if matters weren't tough enough, a blizzard came boiling down in a white swirl of numbness. By the end of five horrible days, four people were nothing but frost-rimmed corpses while another was hopelessly insane and two more had turned back. The remaining ten were not crazy, however, just hungry. With sideways looks at each other, they began to butcher and eat the raw flesh of the bodies.

Several days later, despite the nourishment, the recurrent storms drove another man mad and he shot the two loyal, friendly Indians. At least, the bodies were not wasted. Thirty-three days after leaving the eastern side of the summit, seven of the original seventeen, oddly one man, one boy and five women made it to an Indian winter camp and were saved. Relief parties sent back in the spring discovered with gagging horror how those who had stayed at the lake also made their way through the winter, eating the dead. But, there was not so much to go around and several people died even as they were being carried out to safety. Still, despite the dark events of the legend that will always be the Donner Party, about half the travelers survived and reached California where they found it far more sensible to fade into obscurity that would permit them to avoid the life-long appellage "cannibal."

Under the eat-man-or-die circumstances of the Donner Party, the election to turn cannibal is quite easily understood even by anybody who has ever tried a stiff diet to lose weight. It has many parallels as a tragic circumstance, cropping up repeatedly during Napoleon's retreat from Russia to the grounding of the French frigate *La Méduse* only four years later off Senegal. This was practically an aquatic version of the Donner Party, if even bloodier.

There are, it would only be fair to note, always a few examples of behavior that don't quite fit the handy niches and pigeon holes that house the majority of a subject as broad as human cannibalism. One, which personally gives me more creeps than just about any other aspect of the matter, is that there are some women who, in their odd cravings during pregnancy, get irrepressible desires for human flesh. I suppose, in legal terms, this would be a technical temporary insanity. Unfortunately, it's not as rare as our sense of delicacy might lead us to presume. . . .

Another mighty weird cannibal phenomena occurs among some northern American Indian tribes, especially the Ojibwas and Crees of, mostly, Canada. This disorder is called, in their tongue, *wiitiko,* and it's a cultural psychosis peculiar to these peoples. It begins with an acute melancholia and depression and deepens into a regular killing of men and cannibalism. I guess this falls under What Every Young Brave Should Know about sharing his winter quarters.

Perhaps, as an ex-stockbroker, it's only natural that I would find the greatest interest in cannibalism that has a commercial bend; filthy lucre and all that sort of thing. Well, there are two wonderful examples that typify this sort of situation, one of the best known being the case Fritz Haarmann, fondly known to the press as the "Hanover Vampire." It's not that poor Fritz was really a bad guy, it's just that he was convicted in 1924 of having *bitten* 27 men to death and having made sausages of their meat. Now, that's what I call dedication! If you're going to be a cannibal, at least do it right!

Germany was still on her knees from the Great War of Betrayal when Fritz, 45 years old in 1918, began a whistle-stop tour of Hanover's jails for a variety of petty offenses, usually either theft of one form or another or sexual misorientation. When, after a longer jail term than usual, he opened a restaurant which always seemed to have a good supply of hard-to-get meat, nobody apparently wondered where he got his stocks. While running the restaurant, good old Fritz picked up a few marks as a police informant where he learned the officious technique he needed to demand some runaway's papers, acting the part of an "undercover" policeman, taking pity on the young man and bringing him home for a homosexual frolic, a bite in the throat and a bit of sausage-making. When the parents of one missing young man traced their son to Haarmann and called the police, the cops barged in to find Fritz *in delecto flagrante* with another man and promptly arranged passage back to the slammer for nine months on a charge of homosexuality. Had they not been so shocked by what they walked in on, they would have found the severed head of the parent's missing son wrapped in the newspaper in the kitchen. Presumably, it was pretty ripe by the time Fritz got out of jail!

It was nearly 1920 when Fritz met his guiding angel, sort of a criminal manager named Hans Grans who worked out a reciprocal arrangement with Haarmann in which Grans got the victim's clothes and Fritz got the body. After Fritz bit the young men to death, Hans would strip them and help Fritz butcher the bodies. What was not sold or made up into sausages was eaten by Fritz himself, there being no evidence I can find that Grans was a cannibal. Speaks well of him, no? The larger bones, such as skulls and thighs were kept in a cupboard until they could be thrown out the back window into the Leine River, which was directly behind the Haarmann shop and apartment; not too clever as, after five years, they had gotten so thick that they were discovered and both Fritz and Hans were again nabbed. Fritz, more's the waste, was executed, but Hans got life and was paroled in 12 years. Nobody ever figured out how many neighborhood customers were unwilling cannibals themselves. Fritz was executed on the conviction of murdering 27 people, but the guesses at the time ran between 50 and 150 victims.

Reay Tannahill, in her definitive work, *FLESH AND BLOOD* (Stein and Day, Briarcliff Manor, New York, 1975) mentions another case in Germany in which a butcher had a sausage stand at the main railway station in Berlin where he made quite a nice living from selling peoplewurst.

Of course, as may be expected, World War II is full of cannibalism from the siege of Stalingrad to the prison camps, concentration camps and other high-pressure points where people were forced to eat each other to survive.

You know, researching a chapter like this brings some very weird facts and oddenda to mind. For example, for years I had been eating tartar sauce without ever giving a thought to where the term might have come from. That of course, was before I read about Vlad The Fifth of Walachia, better known as Vlad the Impaler who ruled in eastern Europe in the Fifteenth Century. Vlad had a terrible time with Turks, with whom he was usually at official odds. A second degree cannibal of considerable inventiveness, Vlad once put down a rebellion by the *boyars,* feudal nobility under him, by looting the church of and burning the town of Kronstadt, in (where else?) Transylvania, his grand finale being the next morning when anybody who struck his fancy was impaled on sharp sticks and poles and left to die while husbands and wives were alternately encouraged to eat each other. Some nobility who must have irked Vlad considerably were beheaded, the flesh from their skulls fed to crabs, and the crabs then eaten at a great feast by the noblemen's friends and families. I don't think I would have lent Vlad any money . . .

Some time after this incident, Vlad had a visit from about 300 Tartars and immediately saw them as potential allies against the hated Turks. In a burst of inspiration, he had three of the best Tartars killed and fried, forcing the other roughly 297 to eat them. When they had finished, Vlad advised them that this would go on, one eating the next until whomever was left went to fight the Turks on Vlad's behalf. No dopes, the Tartars hied themselves off to do in the Turks, presumably having witnessed the first serving of Tartar Sauce.

I think, on that somewhat quavering note, we may consider our inspection tour of matters anthropophagous if not complete, at least at an end. Most writers, having packed so much of themselves into a work tend to say something profound at the end of the last chapter of the book, at about the point we now find ourselves. My wife advises me that I have never said anything profound; most certainly not in print. So, I think I'll just quote the obvious message of this book through the untonsured lips of a chap known in his time as Hugh Glass. You'll remember him better as Richard Harris in *MAN IN THE WILDERNESS*. The real Hugh Glass was abandoned in a comotose state at night in the middle of a hostile, Indian-swarming wood on the upper Missouri in the early 1800's by two volunteers, one an 18-year-old the world would remember decades later as Jim Bridger, among other sins the discoverer of Yellowstone Park. Horribly mangled by a wounded she-grizzly bear, unarmed, unable to walk through the territory of some of the continent's most savage tribes, Glass mostly crawled out, through better than 200 miles of some of the roughest country on earth.

Glass, who made it "out" let the revenge that drove him die. He lived a relatively normal life for a few more years until caught on the open river ice by a tribe that didn't like the way he parted his coonskin cap. He was arrowed to death. During his lifetime, he had always been distinguished by a tale he told of himself in which, in ignorance, he had eaten a bite of human stew. A "one-bite cannibal," he called himself.

"What was it like, Hoss," one of his *compadres* once asked him.

The grizzled man who had once lived in Pennsylvania looked into the buffalo chip campfire, pursed his lip and spat a long, amber stream that caught the light.

"Shee-it," pronounced Hugh Glass, Indian fighter, trapper, mountain man and one-bite cannibal: "Meat's meat."

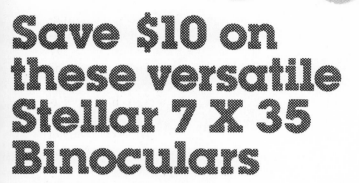